Synthesis Lectures on Engineering, Science, and Technology

The focus of this series is general topics, and applications about, and for, engineers and scientists on a wide array of applications, methods and advances. Most titles cover subjects such as professional development, education, and study skills, as well as basic introductory undergraduate material and other topics appropriate for a broader and less technical audience.

Prafful Mishra

A Guide to Implementing MLOps

From Data to Operations

 Springer

Prafful Mishra
Stockholm, Sweden

ISSN 2690-0300   ISSN 2690-0327 (electronic)
Synthesis Lectures on Engineering, Science, and Technology
ISBN 978-3-031-82009-0   ISBN 978-3-031-82010-6 (eBook)
https://doi.org/10.1007/978-3-031-82010-6

To those passionate about driving change, who push the boundaries of technology every day and make impactful contributions towards engineering at scale.

While working in this space, I realized that a comprehensive guide on scaling MLOps practices was missing. No single resource covered the requirements, challenges, and best practices that are needed to establish MLOps at scale. This inspired me to start this book, hoping it will serve as a foundational guide and a practical companion for all involved in the journey.

To my colleagues, mentors, and everyone who's been part of this adventure, thank you for your inspiration, guidance, and unwavering support.

Preface

Over the past decade, machine learning has come a long way, with organisations of all sizes exploring its potential to extract valuable insights from data. However, despite the promise of machine learning, many organisations need help deploying and managing machine learning models in production. This is where MLOps comes in.

MLOps, or machine learning operations, is an emerging field that focuses on the deployment, management, and monitoring of machine learning models in production environments. This combines the principles of DevOps with the unique requirements of machine learning, enabling organizations to build and deploy models at scale while maintaining high levels of reliability and accuracy.

This book is a comprehensive guide to MLOps, providing readers with a deep understanding of the principles, best practices, and details on how to implement it at scale. From training models to deploying them in production, the book covers all aspects of the MLOps process, providing readers with the knowledge and tools they need to implement MLOps in their organizations.

The book is aimed at data scientists, machine learning engineers, and IT professionals who are interested in deploying and maintaining machine learning models. It assumes a basic understanding of machine learning concepts and programming, but no prior knowledge of MLOps is required.

Whether you're just getting started with MLOps or looking to enhance your existing knowledge, this book is an essential resource for anyone interested in scaling machine learning in production.

Depending on where you stand on the MLOps spectrum, you might find specific chapters of this book of extreme interest.

We are moving in increasing order on the spectrum from *What the F is MLOps* to *MLOps Ninja*.

- *My model has an f score of 0.05. I wanna put it into production. But What is mlops?*—You, my friend, shall start from Chap. 1 (hush, hush, you'll get there soon)
- *Don't tell me fluff about MLOps. I wanna implement it*—Chap. 2 onwards, you go.
- *I got this handled, just tell me what I can do better*—Sorting Hat says Chap. 3 it is.

But no matter where you are, going through the complete journey would lead you to learn some things that you might have missed for sure.

Stockholm, Sweden   Prafful Mishra

Acknowledgments

This book would not have been a success story without the unwavering support from all the people around me. From colleagues and professionals that played an important part in shaping my opinions and mindset, to all my friends that made sure I was able to be in the right state of mind to create this work.

A special thanks goes to the awesome set of engineers I've had the privilege of working with. You've made sure that I was challenged, and that we dive into the complexities of MLOps that are not visible to the common eye. Also, to the people at my work space who not only made my task of managing day to day of this project with work a reality, but also kept me motivated along the journey.

Not to forget the friends that made sure I got the necessary breaks and took care of my physical and mental health along the journey of the project.

Finally, I want to thank you, the reader. It's a privilege to share this knowledge and experience with you, and I hope it helps you navigate and shape your own journey in the ever-evolving world of MLOps.

Contents

Understanding MLOps

Introduction to MLOps

Hello reader! Have you ever heard of MLOps? It's an exciting field that's all about managing machine learning models in production, and with LLMs becoming one of the mainstream technologies in our world, MLOps has also seen a rise in being a demanded skill by employers.

You know as well as I do that deploying and managing software applications can be a complex process, and machine learning is no exception, in fact when ML/AI comes in, this gets crazier with respect to not only scale but compliance and complexity. That's where MLOps comes in (Fig. 1.1).

MLOps is a set of best practices that combines DevOps principles with machine learning workflows. It helps organisations build, deploy, and manage their machine learning models at scale, while ensuring they're accurate and reliable. While making sure that the ML process is flexible enough to rollback and automatically update models in production. Not to forget, these practices help an organisation to maintain and deploy ML based applications at scale.

As a software engineer, you've probably heard of companies like Uber, Airbnb, and Netflix using machine learning to power their platforms. But did you know that these companies also use MLOps to manage their models in production?

For example, Uber uses MLOps to manage their surge pricing model and route optimization model, while Airbnb uses it to optimise prices and attract more bookings for hosts. But it's not just tech companies that are using MLOps. Banks and financial institutions are using it to improve their fraud detection algorithms, while healthcare companies are using it to deploy more accurate disease prediction models.

The benefits of MLOps are numerous. By implementing MLOps best practices, organizations can reduce the time to market for their machine learning applications, increase the

© The Author(s), under exclusive license to Springer Nature Switzerland AG 2025
P. Mishra, *A Guide to Implementing MLOps*, Synthesis Lectures on Engineering,
Science, and Technology, https://doi.org/10.1007/978-3-031-82010-6_1

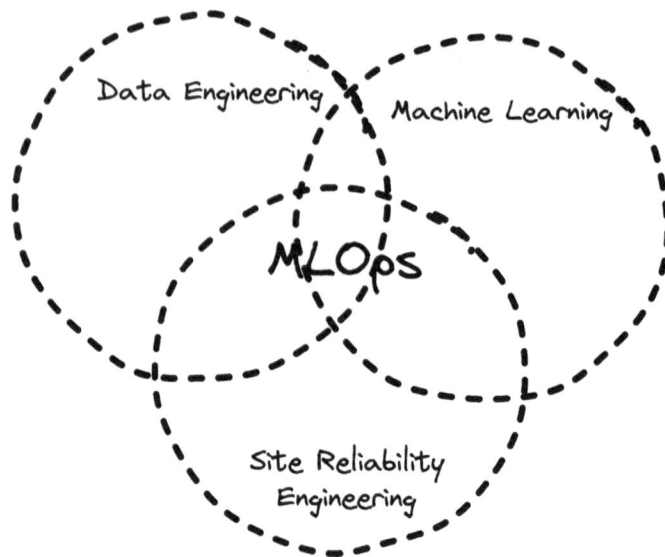

Fig. 1.1 Introduction to MLOps

reliability and accuracy of their models, and improve the scalability and maintainability of their machine learning infrastructure.

By understanding the principles of MLOps, you'll be better equipped to work with machine learning teams and help deploy their models into production.

Challenges of Deploying and Managing Machine Learning Models in Production

Now that you know what MLOps is (I'm sure it sounded pretty simple, right?), let's delve deeper into why it's so difficult to manage and maintain machine learning models in production.

This is a well known fact by now that companies fail to deploy most of their machine learning models to production and even if models make it production they fail to deliver long term value without considerable maintenance and updates. That's right - out of every 100 models built, only 13 actually make it to production. So, why is that? What makes deploying and maintaining machine learning models so challenging?

There are several factors at play here. First and foremost, machine learning models are highly complex. They require large amounts of data, sophisticated algorithms, and specialized hardware to train and run.

Secondly, machine learning models are not static. They need constant updates and retraining to stay accurate and relevant. This means that one needs to have robust processes in place for version control, testing, and validation to ensure that their models are working as intended.

Another challenge is the lack of standardization in the machine learning industry. There are a plethora of frameworks, libraries, and tools to choose from, each with their own strengths and weaknesses. This can make it difficult for organizations to build consistent and scalable machine learning infrastructure.

There's another issue of organizational silos. Machine learning models require input and collaboration from a wide range of stakeholders, including data scientists, ML Engineers, Software engineers, Infrastructure Engineers, and business analysts. Without proper communication and collaboration, it can be difficult to get models into production and ensure they're delivering value to the business, and will continue to do so over long periods with proper maintenance.

These are just some of the challenges organisations face. In the next section, we'll explore how MLOps can help address these challenges and ensure your machine learning models are delivering maximum value to your organisation.

In addition to the technical challenges of managing and maintaining machine learning models, there's also a *people aspect* to consider.

To be successful in the field of MLOps, you need to have a combination of software engineering skills, mathematical skills, and Devops skills (or shall I call it Site Reliability Engineering Skills, *thanks Google*). This means being able to write high-quality, maintainable code, while also having a good understanding of statistical modelling and algorithms and having a deep insight into how these models go to production and sustain a life there.

Unfortunately, many organizations tend to focus on one aspect over the other. Data science teams may excel at building accurate models but lack the software engineering skills to deploy and maintain them at scale. On the other hand, software engineering teams may be experts at building scalable systems but need more mathematical expertise to build accurate machine learning models. Sometimes, both these aspects are catered for, but when it comes to consistently taking models to production and maintaining a lifecycle for deployments, Infrastructure teams fail to understand (or estimate) the needs of ML teams and workloads. I highly suggest reading Cracking the Code of Data Science Team Structures | by Prafful Mishra[1] to understand how one can form a good team for ML and data science.

This is where MLOps comes in. By bringing together the best practices of DevOps and machine learning workflows, MLOps helps bridge the gap between software engineering and data science. It enables organisations to build and deploy machine learning models at scale, while ensuring they're accurate, reliable, and maintainable.

---

[1] [1].

But to truly succeed in the field of MLOps, you need to have a team that's well-versed in both software engineering and mathematics. This means having data scientists who can write production-ready code, software engineers who understand the fundamentals of statistical modeling, project managers who can effectively communicate across these different disciplines, and Infrastructure superheroes (read SREs) who can make the path to production really smooth for all kinds of ML deployments (*we'll go through different kinds of ML deployments later in Chap. 2: Model Deployment*).

The good news is that there's a growing demand for professionals with these skills and will continue to grow as AI creeps in every sector of human society going forward.

How Does MLOps Address These Challenges?

Now that we have talked about the growing demand for MLOps professionals and the challenges they face in managing and maintaining machine learning models, let's take a look at how these professionals can help address these issues. With the unique combination of software engineering understanding of ML, and SRE skills, MLOps professionals are ideally positioned to tackle the complex and rapidly evolving field of machine learning operations.

Let's explore some of the key strategies and best practices that MLOps professionals can use to ensure that machine learning models are accurate, reliable, and maintainable. Whether you're a data scientist, software engineer, or other stakeholder involved in the development and deployment of machine learning models, understanding the role of MLOps is essential for success in this exciting and fast-paced field.

Collaboration: MLOps enables collaboration between all the stakeholders involved in the development and deployment of machine learning models. By working together in a shared environment, these teams can ensure that models are being built with production in mind from the start. This can help prevent issues that might pop up later on, such as difficulties with deployment or maintenance.

Version Control: MLOps provides version control for machine learning models, just as software engineering teams use version control for code. This enables teams to track changes to models, compare versions, maintain compliance, and roll back changes if necessary. This can help prevent issues that may arise when models are modified or updated.

Automation: MLOps automates many of the tedious and repetitive tasks involved in managing and maintaining machine learning models (tasks like data preprocessing, model training, deployment, re-training and maintenance). By automating these tasks, MLOps helps teams save time and reduce the risk of human error, and make Machine learning practitioners more productive.

Monitoring: MLOps provides continuous monitoring of machine learning models in production. This enables teams to quickly detect and respond to issues that may arise, such as model drift or degraded performance. By monitoring models in real-time, teams can ensure that models are performing as expected and make adjustments as needed. This would ensure that the ML products available to users are in the best shape possible round all the time.

Scalability: MLOps helps teams scale their machine learning models to meet changing business needs. This includes scaling models vertically to handle larger datasets, as well as scaling models horizontally to handle more requests or users. By ensuring that models are designed with scalability in mind from the start, MLOps helps teams avoid issues that may arise as models grow in size and complexity.

Robustness: MLops provides the infrastructural best practices to ML model deployments, making sure the ML systems are resilient to failures and are robust. This leads to less downtime, and issues when you ship ML.

MLOps plays a critical role in ensuring that ML is a success story for any organization. By adopting such practices, teams can build, deploy, and maintain machine learning models with confidence and deliver real value to their organizations.

The Business Value of MLOps

MLOps is not just for data scientists and software engineers. It helps business leaders and decision-makers as well. By adopting MLOps best practices, organizations can see some serious benefits that go way beyond the tech team.

Picture this: MLOps can help streamline the machine learning development process from start to finish. That means faster turnaround times and quicker time-to-market for your models, which is a win–win for everyone involved.

That's not all, MLOps also makes sure that your models are accurate and reliable, which is critical for making sound business decisions. Plus, by optimizing your models for performance and scalability, you'll be able to squeeze every last drop of your ROI in machine learning.

And speaking of investments, MLOps can also help reduce the risks associated with machine learning models. With version control, automated testing, and continuous monitoring, you can make sure that your ML products are the best, as you'd be able to spot issues before they become major problems, minimizing the risk of reputational damage or regulatory fines. Additionally, MLOps is a key enabler for the reusability of ML assets, which would mean less money is needed to develop and foster Machine Learning Projects.

Finally, MLOps enables organisations to be responsive to changing business needs and market conditions as new ML technologies come into existence every day. By rapidly developing, deploying, and updating machine learning models, organisations can stay ahead of the curve and utilise MLOps to have the first and most reliable ML solutions in the market.

So, if you're looking to harness the power of machine learning and get real value for your organization, MLOps is the way to go!

Current State of MLOps

Now that we've covered why MLOps is important and the challenges it addresses, let's take a look at what's happening in the world of MLOps today.

The MLOps market is on the rise, and for good reason. With the increasing popularity of machine learning, MLOps is becoming a necessity for organizations that want to make the most of their models.

There are now various MLOps tools and platforms available in the market that can help organizations implement MLOps best practices. These tools offer features like model versioning, automated testing, and continuous integration and deployment, making it easier to build and manage scalable ML systems.

Cloud providers are also offering MLOps solutions that can help streamline the process even further. Platforms provide pre-built pipelines, integrated tools, and infrastructure management, making it simpler for organizations to implement MLOps best practices without having to worry about the underlying infrastructure. There are smaller players already in the market offering end to end MLOps SAAS platforms.

Also, the open source community has been very active in developing and maintaining MLOps related tools. There are a number of open-source tools and platforms available for MLOps practitioners to use, which can make it easier to build, deploy, and manage machine learning systems.

One popular open-source platform is Kubeflow,[2] which is designed to simplify the process of building and deploying machine learning workflows on Kubernetes.[3] Kubeflow provides a range of tools and components, including Jupyter notebooks, TensorFlow training jobs, and serving APIs (now called Kserve), which can be easily configured and customized to suit your needs. Kubeflow also includes support for distributed training and model tuning, making it a powerful platform for scaling machine learning workflows.

Another popular open-source tool is MLflow,[4] which is designed to simplify the process of managing machine learning models across the entire lifecycle. MLflow provides a range of features, including model tracking, experiment management, and model deployment, which can help you keep track of your models and ensure they are performing as expected. MLflow also includes support for a wide range of machine learning frameworks, making it easy to integrate with your existing workflows.

---

[2] [2].
[3] [3].
[4] [4].

Apart from the tools mentioned above, other really good open-source tools and platforms available around the ML space are TensorFlow Extended (TFX),[5] Apache Airflow,[6] and DVC,[7] LakeFS[8], to name just a few. These tools are typically well-documented and easy to use out of the box, making it possible for even novice practitioners to get started with MLOps quickly.

However, even with the availability of these tools and platforms, implementing MLOps at scale can still be a big challenge. A combination of technical expertise, domain knowledge, and cultural change is required to successfully use the full potential of MLOps. However, as the market continues to grow and more organizations adopt MLOps, we can expect to see even more innovations and advancements in this field. Nevertheless, the availability of these open-source tools and platforms is helping to make MLOps more accessible and approachable for practitioners of all levels.

Understanding the need of the new tech era and their ML needs, many companies have recognized the importance of having their own in-house platforms and tools to manage their machine learning workflows. These platforms can be customized to meet the specific needs of the organization and can provide a range of features to support the entire lifecycle of machine learning models.

There are many such examples, where big tech giants have developed internal platforms, which are used to build and deploy machine learning models for a wide range of applications, including fraud detection, forecasting, and personalized recommendations. Such platforms include a range of tools and components, including data preprocessing and feature engineering, model training and evaluation, and model serving and monitoring.

Other companies with their own in-house MLOps platforms like Metaflow,[9] TFX,[10] Abakus (an in-house ML platform at one of the leading automotive manufacturers I used to work at), are few such examples.

Having an in-house MLOps platform can provide a number of benefits for companies, including increased control over the machine learning workflow, better integration with existing systems and workflows, and improved collaboration and communication among teams. However, developing and maintaining such a platform can be a significant investment of time and resources and may not be feasible for all organizations.

---

[5] [5].
[6] [6].
[7] [7].
[8] [8].
[9] [9].
[10] [5].

References

1. https://mishraprafful.medium.com/cracking-the-code-of-data-science-team-structures-c0f754
   dc381e
2. https://github.com/kubeflow/kubeflow
3. https://github.com/kubernetes/kubernetes
4. https://github.com/mlflow/mlflow
5. https://github.com/tensorflow/tfx
6. https://www.github.com/apache/airflow/
7. https://github.com/iterative/dvc
8. https://github.com/treeverse/lakefs
9. https://github.com/Netflix/metaflow

Providing Practical Guidance

2

Introduction

We need to provide practical guidance on how to enforce MLOps in real-world scenarios, and that's exactly what we'll be talking about in this chapter.

When I am presenting at any event about ML/MLOps usually 8 out of 10 people report that their companies are using generative AI. Many of those AI adopters are still in the early stages and are trying to understand how to git generative AI in their existing feature set of the product.

Deploying and maintaining machine learning models can be a complex and challenging process. We know that a lot of organizations heavily invest in having an ML/AI feature and team today, but not all of them succeed in having a stable in house developed product, and have to make do with calling an API of an AI giant organization. One of the main reasons for this gap is the need for proper MLOps infrastructure and processes.

Since the onset of newer AI advancements most of the organizations have increased their budgets for AI and the average number of ML Engineers employed has started going up, and more and more jobs are being available for such positions. However, those same organizations are struggling to manage and scale those efforts. At a lot of organizations the time required to deploy an ML model has actually increased, implying that many organizations are manually scaling their ML efforts rather than focusing on the underlying operational efficiencies that enable businesses to achieve greater results through ML.

TBH, even though a lot of new tech has come into existence, and these advances with Large Language Models/diffusion models[1]/video generation/image generation and audio generation models are painting a good picture of the future (*Yes, this is a pun*) when it comes to taking models to production, a surprising number of companies fail to perform

---

[1] [1].

© The Author(s), under exclusive license to Springer Nature Switzerland AG 2025
P. Mishra, *A Guide to Implementing MLOps*, Synthesis Lectures on Engineering, Science, and Technology, https://doi.org/10.1007/978-3-031-82010-6_2

consistently due to similar challenges as mentioned above. To address these challenges, MLOps has emerged as a set of best practices and tools for managing the machine learning lifecycle. Having said this, the lifecycle management of an ML product spans across a very broad scope, which includes a lot of the following horizontal aspects:

Data is the source of all intelligence. It needs to be treated as a sacred entity. We all have heard *"Bad data In, Bad data Out (GIGO)"*,[2] and a lot has been said about this. From the perspective of ML, one would want to create a lot of versions of data, iterate over different combinations of features, and be able to take one or a few of those combinations to production with ease toward training models.

Model training is a crucial part of this process, and there are many open-source tools available for training machine learning models, such as TensorFlow and PyTorch. Additionally, cloud providers like Amazon Web Services, Google Cloud Platform, and Microsoft Azure offer managed machine learning services that can simplify the model training process.

Once the model is trained, it needs to be deployed in a development, staging, or production environment. This can involve setting up infrastructure, scaling resources, and establishing mechanisms for handling requests and responses. For most of the cloud providers, AI/ML is a top workload requirement driving multi-cloud deployments.

After deployment, it's important to monitor the model's performance over time to ensure it's working as intended. This can involve tracking classic monitoring metrics and ML metrics for model evaluation as well. Mostly in today's k8s heavy world, using tools like Grafana or Prometheus for such tasks has become a de facto norm. This can very well be understood by the number of Cloud Native[3] Solutions for ML available for choice today.

Finally, model maintenance is an ongoing process that involves updating and improving the model over time. This can involve retraining the model with new data, updating the model architecture or algorithms, and incorporating feedback from users or stakeholders.

By implementing best practices and leveraging the right tools, businesses can effectively manage the machine learning lifecycle and ensure their models are delivering real value. In the next sections of the book, we'll dive deeper into each of these components and provide practical guidance for implementing MLOps effectively.

Let's try to delve into these horizontals one by one in a while.

---

[2] [2].
[3] [3].

What is an ML System?

Before we start delving deeper into any specifics, let's discuss a bit about 'What is an ML system?'. You'd find this word used throughout the book while describing a lot of vague things, and this word would be overloaded with respect to this book and real-world definition as well.

Let me try to define what an ML system means as far as the usage is considered in this book.

The complete infrastructure (ranging from data to deployments), culture, processes, and ways of working that lead an ML project from just an idea to a fully working, auto-scaled asset, generating predictions, combine to form this entity called the ML System.

Most of the text of the upcoming chapters is made to provoke thoughts and force you to think about practical aspects of MLOps. When you are doing this mental heavy lifting, make sure to keep this definition of ML System at the back.

ML Project Lifecycle

One the core problems I've seen with Machine Learning teams is the lack of structure and the variance in understanding of how to complete a ML project. There is no correct answer to what is the lifecycle of a project, but its is very important to have a common understanding of the lifecycle by all the people working together.

This can be one of the first standard foundations you as an MLOps Engineer set in place. It needs not be the best or the most accurate description of an ML project's lifecycle. It has to be something that all the ML practitioners that you work with agree to, and it has to be thought of as a live entity, a document that is the preamble to your constitution. It is a document to iterate on as the maturity of your ML system moves towards more complex solutions. 'What stage is your machine learning project in?'is one of the most important question to ask all the time about an ML project, before making any decision. Depending on what stage is the project in, the expectations from the project would be very different.

You can find an example of a ML project lifecycle below, this lifecycle framework is made to serve as a starting point for any generic ML project. For the sake of naming this framework let's call this the **Helix ML lifecycle framework**. The framework defines certain automations that are expected to support each stage of the project (Fig. 2.1).

The stage of a project would be recognised as the value that is associated with the metadata of the project, potentially be stored in the *.mlops* (more about this file in Chap. 3) file at the root of the repository. Changing this variable in the repository would lead to automations behaving differently for the project repository. (Check *Automations* section for each stage to understand behaviour corresponding to each stage).

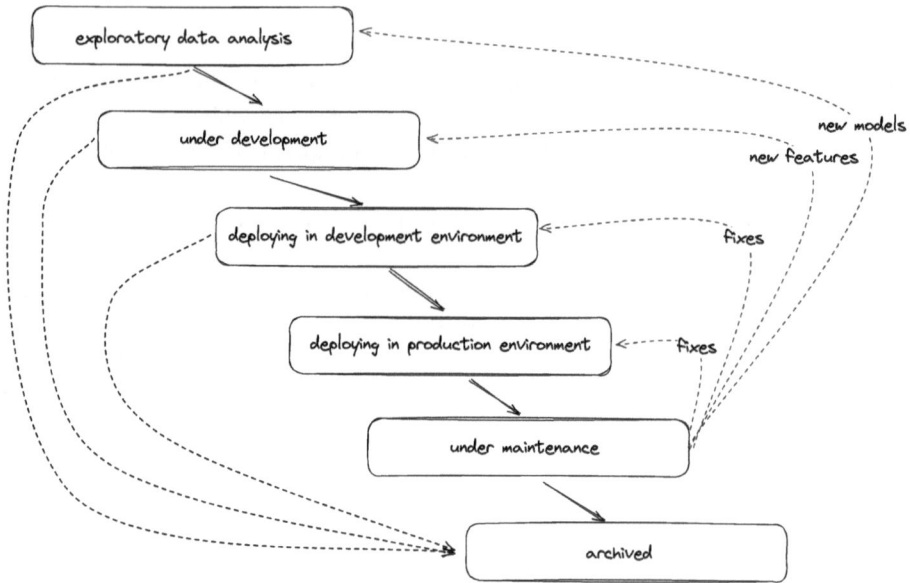

Fig. 2.1 ML Project Lifecycle

Exploratory Data Analysis

This stage of the project describes the time when a proof of concept is in progress, while we are still making sure that the data we have is going to work for creating a solution that solves the problem.

At this stage, we are still not sure if there's a clear solution to the problem and the MLEs should have the maximum possible freedom to facilitate the fastest iteration in this stage.

Expectations

- You are expected to have a git committed notebook, periodically synced at a time schedule of your choice (suggested to be done at least daily).

Automation

- notebooks/ directory in your project repository.
- make command to help with syncing the notebook folder.

Under Development

This stage of the project corresponds to the time when clarity about the path towards the solution has been achieved. This is usually the stage where you have a few known models that you'd like to try with different hyper parameters and you have an estimated design of the deployment in mind. You are actively trying out different models, and are constantly running training pipelines.

Expectations

- All expectations from previous stages.
- The python package of the repository should have in-development code.
- No ad-hoc work is happening on the notebooks.
- notebooks/ directory of the repository is being used in read-only mode and no new changes are being written.
- Unit tests are being written for the source code of the project.
- Pipelines and components for project workflows are under development.

Automation

- CI automation is in place for:
  - Publishing Python package versions.
  - Building container images from Dockerfile.
  - Compiling, uploading, and running pipelines and components on infrastructure.

Deployed in Development Environment

This stage corresponds to the time of project when a model server is being deployed in the development environment of the infrastructure. A lot of tests w.r.t integrability of the server with the other services in the organization are being carried out. Integration tests are being written down and we are trying to nail down all the edge cases to ensure the robustness of the project server.

Expectations

- All expectations from previous stages.
- Unit tests exist for the existing source code of the server and trainer.
- src/, components/, and pipelines/ directories are being used in read-only mode, and most of the work is going into writing good tests/ and implementing fixes if needed.
- Valid CODE OWNERS exist in repositories.
- All the infrastructure resources have an existing team mentioned as the owner.
- Alerts for failure of servers/pipelines are being worked on.

- Work is in progress to make sure that dashboards exist for all components of the serving architecture.
- Dependency (and sub-dependency) versions of the project are pinned.

Automation

- CI for deploying servers to dev infrastructure exist
- CI checks exist to be able to carry out A/B tests
- CI checks exist for running integration tests
- A production readiness review exists and is being worked on.

Deployed in Production Environment

This stage corresponds to the deployment of the model server in a production environment. If the serving method is a pipelined batch job, this stage corresponds to the job being scheduled automatically and is doing batch predictions in a production environment.

Expectations

- All expectations from previous stages.
- The project has passed a Production Readiness Review.
- Unit and Integration tests exist and are run as CI checks.
- Alerts for failure of servers/pipelines exist.
- All alerts have runbooks and are added as a link to the alert message.
- Dashboards exist for all components of the serving architecture.

Automation

- Automation for directing alerts to relevant channels exists.
- System dashboard for common components exist.

Under Maintenance

This stage corresponds to the model being served in the production environment.

Expectations

- The server is serving the model in a production environment
- If there are any alerts in development or production environments, relevant fixes in code and runbooks are made.

- There is an existing team with clear ownership of the project.
- If there's a need to re-train a new version of the model, it is being worked on, without touching the production environment, and it goes through the life cycle stages again to reach production.
- Any change other than fixes, go through the lifecycle stages again iteratively.

Automation

- Automation from all the previous stages exists.

Archived

This stage corresponds to the server not servicing any traffic, and not being used and there are no immediate plans of re-training a new version of the model for this project.

Expectations

- The code repository has been marked archived, and is in read-only stage now.
- All the stateless infrastructure resources have been removed from the environments (like k8s resources)
- All-access related to the project has been revoked.
- All the stateful resources are left as is for versioning and tracing purposes.

Automation

- None.

Data

Data Preparation

Data preparation is a crucial step in the machine learning pipeline. It involves collecting, cleaning, and transforming raw data to make it ready for analysis and model training. Without proper data preparation, the resulting models will likely be inaccurate and unreliable.

Data preparation is a time-consuming process that requires careful attention to detail. It involves identifying and removing missing data, dealing with outliers, and ensuring that the data is in the correct format for analysis. This process can be particularly challenging when dealing with large volumes of data.

Fortunately, there are several tools available to help streamline the data preparation process. A few of the well-known names are discussed below.

It's worth also noting that data preparation is not just a technical challenge. It also requires collaboration and communication between data scientists, data engineers, and subject matter experts. The success of any machine learning project depends on the quality of the data used to train the models, which is why data preparation is such an important aspect of MLOps (*and this is also the zone where the boundaries of data engineering and ML Operations are a bit unclear, and IMHO they should remain unclear*).

Tools

When it comes to tools for data preparation in MLOps, there are a few different options available.

One popular tool is Apache Spark,[4] which is an open-source distributed computing system that provides an interface for programming entire clusters with implicit data parallelism and fault tolerance. Spark can be used to perform data cleaning, preprocessing, and feature engineering at scale.

Spark has a good record of being reliable and scalable. For higher computing needs of chunking data, sparks remote compute clusters prove to be a very good option, and a lot of Data scientists find it very easy to use Pyspark as part of their Jupyter notebooks (*some pros like to just directly work on data with ipykernels*). Being able to use SQL queries to wrangle data using spark[5] and to get it in a dataframe has to be the other very favorite ingredient here. Like every other software on our dear planet Earth, sparks homepage also mentions how you could seamlessly use it for Data Science and ML (*and they are not lying about it*).

Another tool that is commonly used for data preparation is Apache Airflow,[6] which is an open-source platform to programmatically author, schedule, and monitor workflows. Airflow can be used to orchestrate data pipelines, allowing data to be extracted from various sources, transformed, and loaded into the desired format for use in machine learning models. This has to be another common choice as the central tool for a lot of data engineering tools.

It provides an easy way to define DAGs (*read as pipelines*) and is usually used to create data pipelines that extract, transform, and load data between various sources and destinations. It is also very useful to orchestrate ETL workflow to integrate and clean data from different sources for analysis, create data quality checks and alerts, monitor data quality metrics and trigger alerts for anomalies, and set up batch processing tasks such as scheduling and executing recurring jobs like report generation or data aggregation.

---

[4] [4].

[5] [5].

[6] [6].

Some teams have also used it for machine learning model training and deployment in order to automate model retraining, evaluation, and deployment processes. This works perfectly as the basic needs of ML models and data workflows are the same (*i.e., orchestrating a DAG*). Still, Airflow is not designed to be an ML pipelining tool (more about this in later sections focused on <u>Pipelining Engine</u>). Hence, seamlessly integrating it with model inference frameworks could be considered a bit hackish, but it is definitely a functional option. (*If you don't see your model serving needs to grow more than basic, it might be a good option to talk to the Data Engineering team and start using their Airflow for ML workloads as well*).

DBT[7] (Data Build Tool) is a tool designed to help teams transform and manage data in a more structured and scalable way, particularly in the context of a data warehouse or a similar large-scale storage system.

In a nutshell, DBT allows you to write code that transforms raw data into organized, clean, and usable data inside a data warehouse, similar to how you would write code to organize, clean, and transform any other kind of data or files in a typical software project. Think of DBT as being somewhat similar to a build tool for software, but instead of compiling and organizing code, it's compiling and organizing data.

One of my favourite parts about DBT is that it uses SQL as the main language, so you can use your known language as the point of interaction. It also allows you to organize them logically, making your transformations easier to understand and maintain over time. DBT is one of those tools that is made to handle scale and it gets better as the data grows.

It's a data-as-a-service platform that enables one to have massive data lakes and warehouses, and instead of bringing the data to compute (as it has always been thought of), they have the concept of computing where the data is. When you are talking about a massive scale, it shines the best. Not to forget, they provide elastic computing and enable the smooth path to deploying LLMs as well. But I'm gonna stay on the data platform track (*which they are good at when taking huge scale or data mesh architectures into account*).

How Does This Look for Unstructured Data?

Even though I did mention a few tools above, (and there are a million more on the World Wide Web) most of the data problems in the world are solved for structured/semi-structured data. There is no one to blame; the nature of unstructured data itself makes it a tough nut to crack (*or manage*), and when you add the needs of the ML world on top of it, it becomes Scrat's A-corn.[8]

Obviously, blob storage solutions have been the solution for storing unstructured data at very low costs. To an extent, object versioning, Apache Parquet,[9] etc. have been really useful for accessibility of such data. However, it still doesn't answer the questions around:

---

[7] [7].
[8] [8].
[9] [9].

- Querying blob data (more about this later)
- Being able to easily (or cheaply) generate a holistic view of the unstructured data in your bucket w.r.t other data in your organisation (*unless you maintain a metadata state in some shape or form at the time of ingestion, but hey, such techniques are just hacks to make it work*)
- Being able to compare blob file contents easily and at scale.

One of the Big Questions Here is: Which Tool to Use?

Well, as we are talking about ML Operations in the scope of this book, I think it is best to let the Data Gods at your organization make the decision on what suits them. However, it is very important for an MLOperations Engineer to have a good idea of what tools are available and how things work in the Data Engineering Realm.

Having said that, it is recommended to understand the kind of ML problems your organization is going to work on for next x years, what kinds of data (structured/semi-structured/unstructured) will be needed to solve those problems, what kind of data is being collected today, and what are the alternatives for sourcing ML input data later on. With this understanding, it will be very easy for you to articulate the requirements of what you need from data sources, which the data engineering team should take into account, along with the needs of other data consumers they need to cater to.

Another very important aspect to keep in mind here is compliance. Depending on the kind of business your organization is in, there might be strict constraints on how and where your data could be stored or made available.

There are a lot of other tools that can be used for data preparation which are available as commercial data platforms that provide a user-friendly interface for building and managing data pipelines. Most businesses are trying to tailor their offering towards supporting AI and LLMs and the new future. Such tools can help data scientists and engineers easily perform data cleaning, transformation, and feature engineering tasks without requiring extensive coding knowledge.

It's important to note that the specific tools used for data preparation will depend on the organization's specific needs and preferences. The key is to aid the data engineering teams in choosing tools that can handle large-scale data sets, provide some sense of version and lineage tracking, and integrate well with other tools in the MLOps toolchain.

Data Versioning

Well now that we talked a bit about some of the tools for data engineering available, let's dive a bit into why it matters for us. We wish to strive for a very easy way to produce multiple versions of data (by trying out different types of preprocessing techniques or selecting different features etc.), and being able to traverse through these versions, that an MLE could try out with different kinds of models.

This brings us to data versioning which is something often overlooked but a very important aspect of the complete ML cycle.

For data versioning and lineage tracking, tools like DVC (Data Version Control),[10] LakeFS[11] and Pachyderm[12] can be used. These tools allow for the tracking of changes to data sets over time and provide a mechanism for reproducing experiments with the same input data. Additionally, there are some one stop solutions also available that provide a unified platform for data engineering, machine learning, and analytics that include features for data versioning, data lineage, and automated data preparation as a part of the offering.

One very valid question you might be thinking about is: Why do I need to version data? There are a lot of reasons you should be. The most basic and practical answer would be to understand the data on which the model is trained. When you are doing a lot of parallel trials to figure out the best hyperparameter, data, or model type to work, you need to have a sane set of traceable systems to understand and replicate the exact set of variables that led to a working ML model. Along with this, data and models are continuously evolving, and if you are trying to re-train or trying to make a model better, you want to know what is the exact version of data the previous model was trained on, and that's where good data versioning helps you.

Another aspect that this need is driven from is audit and compliance with respect to legal guardrails and laws in place.

Structured Data

As mentioned previously Versioning for structured data is a solved problem and whatever tools you have in hand, most probably are going to solve it in one form or another. Some of them might be best in this case and some might just do the job. In any of the cases it's not a big issues to handle structured data w.r.t versioning as you could imply simple methods like including timestamp features or simply by creating a version tag and it would just work for you.

Unstructured Data

This is the trouble, the REAL MESSY TROUBLE, and more the data more the mess. Realistically speaking, every big organization is dealing with cleanups of a lot of data mess. As the world has grown to be data-driven, having clean, clear and accessible data at scale is one of the core capabilities of a successful business.

Having said that, unstructured data is at the core of this mess for a lot of businesses, and IMO there is no solution that solves this problem to a 100%. You have to make trade-offs somewhere or the other, and optimizing the least mess is the way to move here (*at least as of today, I hope something new comes up later in life that solves this problem*). Let's look at a few good options for versioning data though.

---

[10] [10].

[11] [11].

[12] [12].

Basic Object Bucket Versioning

One of the simplest working idea for versioning data is directly using the object storage services available with every cloud provider today. Most of these object storages allow you to enable versioning at the object level, and then once can query any version of the object, treating the last saved version as the latest one. This is a very simple and naive way of maintaining multiple versions of objects and treating the latest version as the default one for queries.

One may have to build a few small hacks around this idea and may not get all the fancy and detailed functionalities and integrations that one gets with more mature and better tools. This works for a small scale, and is a super effective way of maintaining versions, tracking and sanity of the assets. Simplicity is the key of this way of versioning. Hence, is still a good consideration for versioning unstructured data.

DVC[13]

DVC tracks versions of files by seamlessly integrating with existing git versioning used for code by the user.

You have to configure an object store bucket with DVC, and DVC acts as layer between the object storage and the user to ensure the objects. Let's go through a basic use case to get a better understanding of how to use DVC.

Once you wish to upload a file to the storage,

- DVC would compute a checksum for the file content and create a .dvc file.
- It generates a path, and uploads your data file to that path in the object storage bucket configured as the storage.
- This .dvc file acts as a replacement for your original data file.
- You are expected to add the .dvc file to your existing git repository with code.

When you wish to download a file,

- DVC reads at the local .dvc file.
- It computes the path of the object in the file storage from the contents of the .dvc file.
- Download the required file and send it to your local system.

An example of a '.dvc' file is as follows:

outs:
    - md5: a304afb96060aad90176268345e10355.
    path: data.xml.
    desc: Cats and dogs dataset.
    remote: myremote.

---

[13] [10].

On top of the above use-case, a local cache is maintained in the local machine as well to ensure that downloading large files is a non-repeated and inexpensive operation.

If you are planning to implement this, I would highly suggest to:

- Create a data directory in your repository and add all the .dvc files to that directory.
- Use one bucket per git repository to have easier management of data.

Moving towards a bit more complicated and advanced stuff that DVC could do for you.

You can create pipelines, and automate data workflows for you as well. There is experimentation management as well that could be utilised (but it is very basic IMHO, especially when compared to other options available today) for simple use cases.

There are a few things that might not play well when using DVC:
One of them is that as your ML people use local machines a lot to get started with their EDA, and to get some ground work before moving to pipelines, the local caches might get very heavy after a while, which is not compatible with development on personal machines after a scale.

If the cardinality of files that are used in a project become huge, having a .dvc file per object is not feasible or maintainable.

Sometimes Data/ML practitioners coming from different backgrounds may not be super handy with handling git, and adding the layer of DVC increases the learning curve for them.

Eventually, making data versioning a more difficult problem to work with.

Sometimes, you might want to version data that is not related to a use-case or git repository, in such cases having to use a git repository just to track data in object storage might just be an overhead.

With all these considerations in mind, it's best to look at tools with their latest offerings into account, and to verify with the use-case before deciding it as the tools of choice in any ML setting.

LakeFS[14]

LakeFS brings a completely different approach to the Data versioning problem, where you get a repository for storing your data and can maintain a git-like branching strategy for your object storage.

This git-like branching model makes your data lake ACID compliant by allowing changes to happen in branches that can be created, merged, and rolled back atomically and instantly. In a nutshell, think of LakeFS as a **git but for your data**.

This can be a nightmare for people coming from academia and research backgrounds, as they could already be struggling with still catching up with git for maintaining their code. But sometimes you have to look at the benefits against the learning curve.

---

[14] [13].

There is a huge support for multiple integrations to be used with the tool, with integrations with tools like Apache Spark, Kubeflow, Cloudera, Vertex AI and many more.

To make it work with git, you could use the following approach, which was used by my team at one of the organisation I used to work at, and was later suggested to lakeFS as a feature (this is implemented and available as a feature today as LakeFS Git Integration[15]).

LakeFS acts like a layer between users and data storage as well, but it is a complete tool in itself rather than using another versioning system (contrary to DVC). The issue here is that having to maintain different repositories for data and code which has its own versioning engine can be a maintenance hassle (Fig. 2.2).

Sometimes you would want to maintain a clear connection between the version of data used and the version of code used to produce assets. To achieve this, we used inspiration from DVC and created a file called .lakefs and used this file to store the commit ID of data in lakefs. We also created some automations around ML projects to make sure upload and download commands take the commit ID in the local .lakefs file.

We added a suggestion to lakefs explaining how we wish to use lakefs to get their thoughts around it.[16] This, however makes the connection between data and code very clear and makes it very easy to track lineage of ML assets, is a half-baked hackish solution as one cannot handle merge conflicts at scale with this, when it comes to merging data files.

As we can observe from the above image, the git trees for both the repositories (data and code) diverge quite a lot after a while. We came up with a request to add merge strategies or rebase options for lakefs branches. This makes sure that as the branch for code is merged with other branches, the data branch also gets merged accordingly.

But this is also a good thought exercise at the level of data versioning and lakeFS as well.

Having an option to use a Lakefs merge strategy[17] is a good option to good walk around the issue (which was a part of the suggestion).

Another interesting challenge here is compliance with GDPRs right to be forgotten,[18] when you are talking about creating and maintaining multiple versions of data. One potential option that discussions with LakeFS team led us to Garbage Collection 2.0 spark jobs,[19] which would traverse all possible commits of the objects on the storage, and remove the file, leading to a tombstone pointer when someone needs to access a deleted file.

Lakefs duckDB Integration is another recent and very powerful feature of DuckDB[20] integration. DuckDB is an in-process SQL OLAP database management system. You can

---

[15] [14].

[16] [15].

[17] [16].

[18] [17].

[19] [18].

[20] [19].

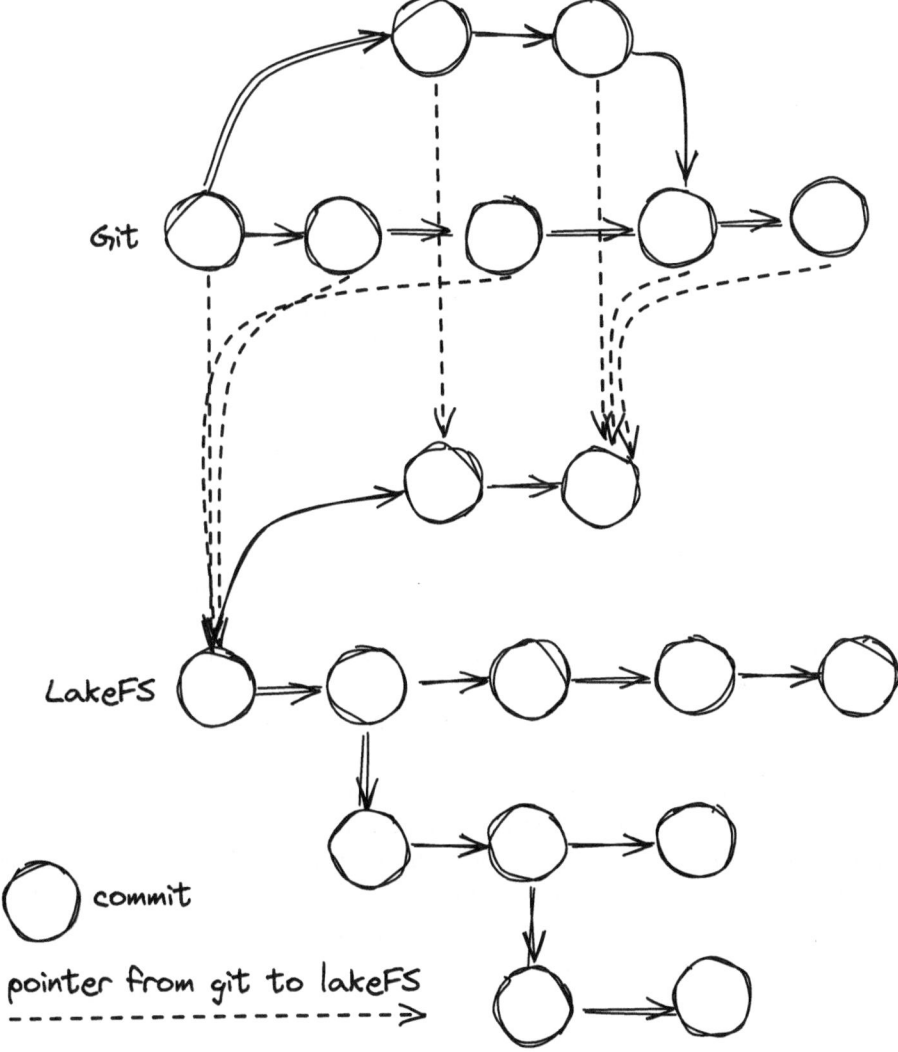

Fig. 2.2 Git connection to lakeFS

access data in lakeFS using a DuckDB layer that you'd interact with or use the inbuilt
DuckDB interface in lakeFS.[21]

---

[21] [20].

Seamless Data Access

Now that we have discussed data versioning, let's just talk about the importance of seamless data access for ML. It is a critical aspect for machine learning (ML) practitioners, as the quality and availability of data play a pivotal role in determining the performance and accuracy of ML models.

Ensuring the reliability and consistency of data is paramount, as inconsistencies or errors in datasets can lead to biassed or unreliable results. Data engineers are responsible for implementing robust data validation and cleaning processes, conducting regular audits, and maintaining high data quality standards over time. *"It can also be fruitful to have some ML or MLOps engineers highly involved in the process of creation of datasets."* This not only ensures there is user perspective involved in creation of datasets, but also makes the user feedback loop very short and small.

Timely and reliable data retrieval is equally essential for ML practitioners, especially when models require real-time or near-real-time access to data for making predictions or adapting to changing conditions. Data engineers must focus on optimizing data retrieval processes, implementing efficient caching mechanisms, and designing data pipelines that enable low-latency access, ensuring that the ML models operate with the latest and most relevant information.

Scalability is a significant consideration as datasets grow in volume over time. To address this, data engineers should design scalable architectures, leverage distributed databases, and employ parallel processing techniques to handle larger datasets efficiently. This ensures that the infrastructure supporting data access can scale to meet the demands of increasing data volumes without compromising performance.

Data security and compliance are crucial aspects of seamless data access. Protecting sensitive information and adhering to regulations is imperative for ethical and legal reasons. Data engineers must implement robust security measures, including encryption and access controls, to safeguard data. Regular audits and updates ensure ongoing compliance with evolving regulatory requirements.

Data integration is another crucial factor for ML practitioners, as models often require data from multiple sources. Seamless integration of diverse datasets is critical for comprehensive analysis. Data engineers play a crucial role in designing and maintaining ETL (Extract, Transform, Load) processes to integrate data from various sources, adapting to data formats and structure changes over time.

Proactive monitoring and logging are essential components of maintaining seamless data access. These measures enable data engineers to track system performance, detect anomalies, and address issues promptly. A well-established monitoring and logging system is instrumental in identifying potential disruptions to data access, allowing for timely intervention and ensuring the continuous reliability of ML models.

As an MLOps person making sure that the Data professionals are aided and there is fluent and smooth communication w.r.t all the above described aspects is very crucial for the success of the ML and Data landscape at an organization.

Data Validation and Monitoring

Data Validation and monitoring is very crucial in ensuring all the aspects described above are kept to the highest standards.

Once a data source has been used to do some exploratory data analysis, constantly monitoring if the data still satisfies the constraints that were assumed while carrying out the modeling is important to make sure that the models created from this data keep performing as they should.

This monitoring needs to be carried out on incremental batches of the data as it might be needed for automated (or manual) retrainings. Also, in the use case of batch predictions it is really important to make sure the prediction inputs are following the standards that the model is expected to perform.

For example, let us assume we are creating a model to generate a representation vector for an audio input, and EDA conducted on the available data for training exposes that all the audio samples we have are only music audio files, and naturally an assumption is made that the prediction input would also be all music. In such a scenario, it might make sense to tune the model to work best with music audio inputs. But it is necessary to put a data validation on prediction inputs that are fed to this model, and to raise an error if the prediction input doesn't contain music, rather than try to create a representation vector for it, which might lead to further worse predictions depending on what the vectors are used for.

Taking the same example to the next stage now. If we wish to refresh the model periodically with an automated re-training pipeline every month as we collect new training samples, it is very important to flag any issues with the data at the time of ingestion itself, making sure there are enough alerts that get triggered once this happens.

Making sure that such alarms and monitoring for validation data and flagging any drifts in the data distributions, are in place is a very good combined project for the Data Engineering and MLOps teams.

Even though such good measures might be in place, it is best to have some small-scale validation in place at the usage layer of all this data. One tool to have such checks in place is called great expectations.[22] As the name suggests, you can have great expectations from the data you wish to consume, and if the data doesn't match the expectations, you can raise an error, process the data in a different way, or take another specific course of action according to your use case. You can define your expectations at various levels like expectations from the batch, table, query, column, etc.

---

[22] [21].

It's a good practice to add great expectations as a part of an automated training pipeline to not train and have an ML Engineer or Data Scientist manually verify the data and perimeters before pushing the final start training button.

Having such data validation, monitoring and alerting systems in place can take the maturity of the ML systems of an organization to a completely new level.

Feature Engineering

Feature Engineering is like adding the secret sauce to your machine learning model. It involves transforming raw data into meaningful features that can help your model make better predictions. For example, if you're building a model to predict housing prices, you might extract features such as the number of bedrooms, the square footage of the house, or the distance to the nearest public transportation. (*Well, we all understand that this is an oversimplification of the process and complexity*).

The challenge with feature engineering is that it can be time-consuming and requires a lot of domain knowledge. Data scientists and engineers need to have a deep understanding of the problem domain to select the right features and create meaningful transformations.

Being able to try different types of features, with different kinds of preprocessing at scale can be really helpful in reducing the speed of iteration for an experiment (or trial as some people prefer to call it) and this is where the context of MLOps comes into picture.

Automating parts of or all of feature engineering can be a game-changer. With the right tools and processes in place, data scientists and engineers can create reusable pipelines that extract, transform, and select features automatically. This can help them reduce the time and effort required for feature engineering and enable them to try a wider spectrum of features before making a final choice on the finalized models.

Tools

There are various tools available to support feature engineering in the context of MLOps. Here are some examples:

Feature tools[23] is a Python library for automated feature engineering. It helps to create features from relational datasets without requiring manual feature engineering. So, if you're working with relational data, you'd have to write code to create features, which can get super tedious. With Feature Tools, you just kinda point it at your data, and it figures out how to create those features automatically. It's pretty handy if you've got a bunch of complex data and don't want to spend forever prepping it.

---

[23] [22].

Google Cloud AutoML Tables[24] is a bit different from this. It's a part of Google Cloud that provides automated feature engineering. What's cool is that it doesn't just handle feature engineering, but the whole machine learning process. So you give it your tabular data (like spreadsheets, database tables, that kind of stuff), and it'll create the features for you and also train and deploy models. It's like an all-in-one system.

TensorFlow Transform,[25] on the other hand, is more focused on data preprocessing, and provides a set of pre-built functions for feature engineering and transformation and also allows users to define their custom functions. So, if you're using TensorFlow already, this makes sure the transformations you apply to your data during training are also applied when your model's live in production. It's got some pre-built functions for transforming your data, but you can also write your own if you need to do something more custom. As this takes the same feature engineering for training and predictions, you don't have to stress over any inconsistencies between both those times.

And then there's PyCaret.[26] It's really built for automation, so a lot of the stuff you'd usually spend time doing manually, like cleaning data, selecting features, or even testing different models, PyCaret handles that for you. It's kind of perfect if you just want to focus on the results and not get bogged down in all the details, especially for quick experimentation. In the MLOps space, tools like PyCaret are gold because they help streamline the process without losing too much flexibility. You still have control, but the routine stuff is taken care of. So, if you're juggling a lot of tasks or you're collaborating with teams that want to see results quickly, PyCaret gives you that speed without sacrificing too much quality.

As usual, the choice of tool is really dependent on what kinds of data needs to be fiddled with for being able to create features. Whether we are talking about images, or audio or we are thinking of only structured data for creating the ML magic.

It is usually better to take the KISS principle,[27] and do not build fancy stuff until absolutely needed. It can be really alluring to build really cool stuff that is very fancy and solves a really good feature engineering problem, but it would be of no use unless it can be put to use.

Easy Integrability

Whatever tool you choose, it's important to choose ones that are easily integrated into your existing workflow. Look for tools that support multiple data formats and allow for efficient feature extraction and transformation. It's also a good idea to choose tools that support feature stores, which can help manage and share features across your organization.

---

[24] [23].
[25] [24].
[26] [25].
[27] [26].

And just like with other MLOps tools, it's important to select user-friendly ones and provide good documentation and support to ensure a smooth user experience for data scientists. Remember that in many cases, you may need to support multiple feature engineering tools to meet the needs of your organization's data science teams. The tool should also be seamlessly integrated to the existing infrastructure in our organization.

But wait..wait....wait........wait, Did i just say Feature Store? That fancy thing that everyone keeps throwing around as a term? What about it? What is it? How do we use it?

Feature Stores

What is a feature store? Is it just a glorified data store? Well, yes and no. (*If you think that a datastore with specific use-case and some added functionalities is a glorified data store, then yes*). But yes, a data store that has specific features that makes it easier for the user to create multiple views, lets them to treat this view as a first class entity, and enables them to serve this view at scale could be termed as a fully functional feature store.

Feast[28] was one of the first feature stores the world ever saw in existence, and since then the ML community has never looked back. A feature store can register different kinds of features, store them and serve them at scale as per one's need.

Feast is an open source tool that could be deployed in house (provided your MLOps team has the bandwidth to handle the maintenance cost) or if you are looking for a more managed solution you could consider Tecton,[29] which is a managed feature platform. Tecton provides much more than the discussed functionalities. For most of the use-cases, feast would suffice as a feature store and server. Tecton could be considered once the data maturity at an organization is at a good scale and their are significant challenges in accessing, transforming and serving features.

Many cloud providers have made a good effort to repackage the existing underlying capabilities to create a feature store like offering (which they call a feature store). But I would say these effort has not been super successful at providing all the features a feature store shall have.

Although one could spin up a real-time feature serving compute machine and use it to train models at almost real-time IO speed between the training compute machine and the online serving machine, it fails to deliver on these functionalities when we talk about unstructured data or other object storages.

Perfect Point of Contact for Data Engineering and ML Data Consumers: One other very common challenge teams face is to understand where to draw the line between a Data Engineering Team and MLOps teams. As discussed before this could be a thin line, and feature stores are exactly on that thin line.

---

[28] [27].
[29] [28].

Some options to consider here are:

- Data Engineer takes the responsibility of populating feature stores from the parent data sources for ML Engineer to consume and MLOps helps MLEs to ensure good practices of using the Feature Store downstream projects.
- MLOps takes the responsibility of populating feature stores from the parent data sources for ML Engineer to consume and helps MLEs to ensure good practices of using the Feature Store downstream projects.

The end decision on which way to move ahead could depend on a lot of specific pointers ranging from expertise of people available, team structures, pattern of data consumption, maturity of data management etc.

EDA

EDA or more commonly known as Exploratory Data analysis is a very important aspect of being able to crack a good ML problem, and this is the space where adding minimum friction and maximum flexibility for an MLE can do wonders.

MLEs having maximum flexibility sometimes are thought to be contradictory w.r.t MLOps principles that point towards robustness, faster to production experiment cycles etc. But If one nails down the perfect balance of control at the EDA stage, which does not hinder the creativity of an ML practitioner, that would be the best case scenario for all the parties involved.

I like to think of this stage of an ML project as the canvas stage of a painting; before EDA, the canvas is still empty. You might have made decisions on what paints you are going to use and what size brushes you are going to draw strokes with; you might have a broad idea of what your painting might look like or what you are going to draw (assume this to be your understanding of the problem and the ideas in your plans of approaching the solution), you might have decided on what canvas material to use, and now you are just about to start with the first strokes of your creative journey. But suddenly, MLOps says, "Oh! Your color palette is not ordered in sequence." well, you don't care at this moment. (*such control freaks these MLOps people sound, I tell you*).

You would not want to disturb the artist at such a vital time, so you should not ask about "oh! you don't have your notebook stubs committed in git". Having said that, once someone has figured out how they wish to move ahead with creating the model, or once they have better evaluated the data available to them, there should be a way for them to easily use the notebooks they created in further work.

This can be achieved using Jupytext maybe (although github has solved the diff problem with ipynb files), but it's still valuable to have good defined functions which could be used in your project's python package using local imports (although this is not a best-practice, but would be a good feature to have). This is illustrated with the ml project template section later under Chap. 3.

Why Are VMs Still Used for This Job?

A lot of MLEs still prefer to start a VM with their choice of their cloud provider, and use it. But this often is not the best way to utilize resources that are tied to the notebook.

This may not seem like a big problem, but one of the core principles of MLOps is "Saving Money" (*Others might say this principles with other names like being resource effective or squeezing all the juice of the lemons, but you get the point*). One should ensure a good way to make sure that EDA happens in the most flexible and resource efficient way.

So the big question here is that Is there a better solution than using bare bone VMs? and the answer is yes.

Most of the notebook solutions available today by almost all the cloud providers have some sort of resource utilisation and auto-shut down built into them. Google colab[30] allows you to detach and attach machines behind a notebook. So do AWS, and others as well.

One major difference is that Google colab has a completely built solution, but other cloud providers are based on providing the good old jupyter notebook as a machine. Kubeflow notebooks[31] is a good alternative as well when it comes to being able to launch on-demand jupyter notebooks powered by pods on kubernetes.[32] Kubeflow also provides a culling mechanism to make sure notebook servers not in use are shut down to avoid resource wastage.

Jupytext[33]

Jupytext is one of the tools that could be used to draw a good center line between the two extremes. It should be used to jupyter notebook to .py files, which should be committed to a git repository. Maybe under a directory called notebooks/ in the repository, to indicate that this code is generated from a notebook, and might not be ready for use in any production setup.

However, if this directory is converted to a module with a small __init__.py file, this could be used to import some functions in the src/ of the repository.

---

[30] [29].
[31] [30].
[32] [31].
[33] [32].

This could be a very good motivation for MLEs or DSs to make cleaner choices when doing EDA in notebooks.

Kubeflow Notebooks[34]

Kubeflow notebooks is an open source tool that could be used to spin up container based environments for EDA using the compute resources of a kubernetes cluster. It defines a Notebook CRD, and one can configure to run any kind of platform of container images[35] (including, but not only the above 2 described solutions). It could easily be used to provide Rust/Golang based EDA environments to MLEs and DSs.

It is designed to simplify the deployment, scaling, and management of machine learning (ML) workflows on Kubernetes. There are a lot (really a LOT, and maybe much more than needed) of similar solutions provided by every MLOps solution today. In my honest opinion if an EDA tool can be easily integrated in the ML projects, is allowing ML practitioners to be flexible, is very non-opinionated in every aspect, and still is able to provide some reproducibility and versioning of code, it is perfect for the job. There's nothing more expected from such tool.

The great part is that the users are not limited by the power of their laptop. They can tap into Kubernetes to scale their resources as needed. So, if they need a GPU to train a deep learning model, it's just a matter of specifying that when launching the notebook. Once they're done experimenting, they can shut down the notebook to free up the cluster's resources.

Plus, since everything is running inside the Kubernetes environment, the transition from experimenting in the notebook to training models at scale is seamless. You can connect this notebook to the rest of the Kubeflow ecosystem, like Pipelines for automation and Model Serving for deploying the model when it's ready.

Code Servers

Code servers[36] have can be another very easy way of providing EDA grounds. Using Code server as container image backed on a scalable compute resource (*something like or exactly kubernetes*) is a very easy way of solving such problems. Code server is essentially a complete IDE, which can be used with any ipykernel[37] based extension, and git to make sure the desired (and above described) behaviour can be replicated.

---

[34] [33].
[35] [34].
[36] [35].
[37] [36].

It is basically a way to run VS Code in your browser. It's like having your coding environment on a remote server, which you can access from anywhere. Imagine you're traveling or using a different machine but still want all your VS Code settings, extensions, and files. With CodeServer, you just log in through a browser and bam!!! You're back in your development environment, no matter where you are.

This also helps in making sure that you are not working in a notebook, but you have your full fledged development environment with you. I've definitely considered using such an IDE experience on devices that just have a browser and dont support for running such applications on the device itself.

Model Development

Model Training

Let's look at the process of model training and the process of deciding on a model to train (EDA) with the following aspects. One might be thinking that wait this is the work of a machine learning engineer how does mlops come into it. Yes, you're right, that makes a lot of sense, but, to enable a machine learning engineer to create production worthy Pipelines, the MLOps engineer should have made a really strong choice that is in accordance with the machine learning problems the organization is solving and would be looking forward to solve in the future.

One of the prominent questions an MLOps Engineer has to answer here is:

Should you be using kubeflow[38]/airflow[39]/mlflow[40]/dataflow[41] or maybe just a container that runs on the infrastructure of your choice?

Of course there might be other options as well, and as time passes by more and more options would exist (But you get the point of the question, *How to manage the infrastructure needed to train models?*).

> Maybe you wish to run training jobs with your on-prem devices, but this conversation is still valid, all thanks to kubernetes.

Kubernetes is a very good solution not just to manage resources and infrastructure but also to remain cloud agnostic. This can be used with simple choices like argo workflows[42] to

---

[38] [37].

[39] [38].

[40] [39].

[41] [40].

[42] [41].

further orchestrate the training workflows into pipelines. Whatever the choice is, it should be taken under consideration of some very important factors:

- What does the core infrastructure use for orchestrating long running workloads?
- What infrastructure the ML practitioners used? How much time would be invested in introducing and adapting something new?
- What is the expected scale (resource utilisation and time) of training jobs ML practitioners are planning to train in the future ?
- What tool would have the easiest DevX for users?
- What is the plan for support available if the training infrastructure has an issue?

If you can get a clear answer to the above questions, the answer for deciding the model training infrastructure would be in front of you.

Packaging Your ML Code

So I'm proposing this ML code packaging strategy that I like to call the "Onion Package". The basic idea here is to create something that wraps all the layers of what you need into a neat package, like how an onion has layers, right? The core of this thing is the Python package—which is the center of everything—then we wrap it in a container image (OCI), and then we put another manifest layer around that. Think of it like layers you peel off when you need to, but it's all bundled together in one unit (Fig. 2.3).

Fig. 2.3 The onion package

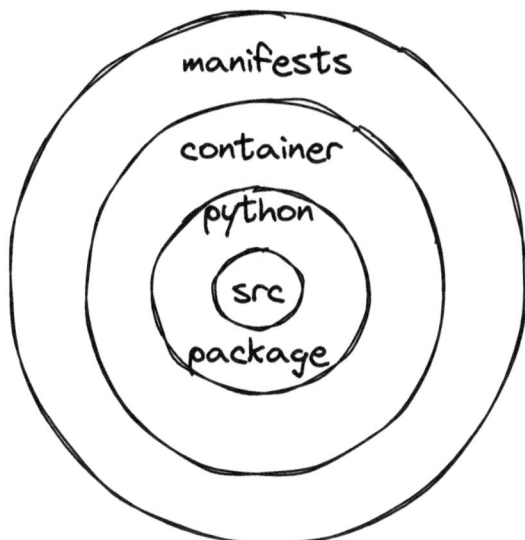

So, the core is your Python package that holds your model code, dependencies, utils—whatever. Then we take that and package it inside an OCI container. We do this because containers give us a consistent environment to run the model in, and they make deployments a lot easier in almost any kind of environment today. We know containers, they help make sure it runs the same everywhere.

Then, on top of that, we add a manifest. This is like the outer layer that holds the metadata—versioning, runtime configs, where does the contianer image live, all that kind of stuff. So, you can keep track of what you're running and easily replicate or upgrade the package without breaking things. The nice thing about this is it keeps everything organized and portable, like you can move your onion package to different environments or share it across teams, and it still works.

The benefit of this? Well, it gives you a consistent way to package and deploy ML models with all their dependencies. Plus, it's got layers of abstraction, making it easy to peel back and debug or optimize. And with OCI, you're using an industry standard that works with any cloud.

It's just like this layered approach—everything you need wrapped in one package that can be deployed and scaled without worrying about compatibility or environment issues. Like, you don't just dump the Python package on its own, you got it covered in layers for stability.

Just one more minor detail, just like when you peel an onion you can choose to remove as many outer layers as you wish and you still can use the onion. You can do the same with this, remove the manifest layer, still a usable container image; remove the container image, still a usable python package left; remove the python package and you can still use the core src/ python module.

Versioning

In MLOps, version control is crucial not only for the code but also for data and artifacts.

Data versioning is important because ML models depend on the quality and relevance of the data. The same model trained on different versions of the data can produce different results, which can have a significant impact on the business. In addition, data sets can be large and complex, with multiple sources, transformations, and pre-processing steps. Keeping track of all these changes can be challenging without a proper versioning system.

Similarly, code versioning is important because it allows teams to collaborate effectively and maintain a history of changes. Code changes can have a ripple effect on other components of the MLOps pipeline, such as data pre-processing, feature engineering, and model training. Keeping track of all the changes and their impact can be difficult without a versioning system. Versioning code also enables reproducibility, as it ensures that the same code is used to train and deploy the model.

Artefact versioning is important because it allows teams to keep track of the different versions of the model, its metadata, and associated files. Artefacts can include the trained model, its configuration, evaluation metrics, and any associated files or dependencies. Keeping track of these artefacts is important for auditing, debugging, and rollback purposes.

The need for versioning in MLOps is even more critical in regulated industries, such as healthcare and finance, where compliance and accountability are paramount. Without proper versioning, it can be difficult to demonstrate the provenance and lineage of the model, which can lead to legal and reputational risks.

Versioning is a critical component of MLOps, as it allows teams to collaborate effectively, maintain a history of changes, ensure reproducibility, and demonstrate compliance. MLOps platforms typically provide versioning capabilities for data, code, and artefacts, making it easier for teams to manage and track changes in the pipeline.

Tools

There are several tools available today that can help with versioning in the context of MLOps. Some notable mentions include:

Git: Git is a widely used version control system that allows teams to collaborate on code and track changes over time. Many MLOps teams use Git to version their code, along with other tools that help to manage dependencies and automate testing.

DVC: Data Version Control (DVC) is an open-source tool that provides version control for data science projects. It allows teams to version data, track experiments, and reproduce results.

Lakefs: lakeFS (as we read in the data sections before) is an open-source platform that provides version control and management for data lakes. It enables teams to treat data as code by providing a Git-like interface for data lake management. With lakeFS, you can version control your data lake's objects (such as Parquet, CSV, and ORC files) and metadata (such as Hive and Glue tables), making it easier to track changes, roll back to previous versions, and collaborate on data lake changes. LakeFS integrates with other data lake tools such as Amazon S3, Azure Data Lake Storage, and Google Cloud Storage.

MLflow: MLflow is an open-source platform for the complete machine learning lifecycle. It provides tools for tracking experiments, packaging code into reproducible runs, and sharing and deploying models.

Kubeflow: Kubeflow is an open-source machine learning platform that is built on top of Kubernetes. It provides a set of tools for deploying, managing, and scaling machine learning workflows, including versioning models and data.

Neptune: Neptune is a cloud-based platform for data science collaboration and experimentation management. It provides version control for data, models, and code, along with tools for tracking experiments and visualising results.

I could write more about what are the pros and cons of using these tools and when to use which tool, but as the devil is always in the details, it depends on your circumstances, needs and existing infrastructural situations. All of these tools help MLOps teams to version their data, code, and artefacts, which is essential for managing the machine learning lifecycle effectively. By versioning their work, teams can easily track changes, collaborate with others, and ensure that they can reproduce results.

Now, let's talk about the people aspect of versioning.

Whaaaat? Does Versioning have a people aspect to it? you must be thinking. Well, let me introduce you to this awesome person called Data Scientist, very well versed with numbers I must say. They can find gold in any data. But sometimes (read often), they don't get along with versioning.

As we all know, most of the data scientists come from a maths background and are more focused on building complex models. They might not realise the importance of versioning data, code, and artefacts. This is where the role of MLOps comes in. MLOps professionals help bridge the gap between data scientists and software engineers, ensuring that versioning is given the importance it deserves.

Data scientists need to understand that versioning is not just about keeping track of changes. It also helps in the reproducibility, collaboration, and sharing of models across teams. With versioning tools like Git, one can easily track changes in code and data, collaborate with team members, and share their work with others.

Similarly, MLOps professionals can help data scientists understand the need for versioning tools like lakeFS, which can help version data on object storage systems like S3, GCS, or Azure Blob Storage. By using tools like these, one can ensure that their data is properly versioned and can be easily reproduced, which can save a lot of time and effort in the long run and helps in recovering assets, when luck is not by your side.

A good and common way to think of ML development is to think of a Chemistry lab, where you are looking to make a new compound, and you'd want to try a good number of ideas, and until you keep a track of what chemicals, how much of each of that chemical, what environmental conditions you are trying out your experiment in, there is no way to reproduce the same experiment again, or to go back to the 7th last thing you tried last week.

Overall, versioning is a critical aspect of MLOps, and both data scientists and software engineers need to understand its importance. By working together and using the right tools, teams can ensure that their experiments are properly versioned, making them easier to manage, maintain, and share.

Pipelining Engine

Oooooo! This almost reminds me of Mario, but let's talk about pipelines. Pipelines are a well utilised concept from the Data Engineering world, and as there is no ML without access to Data, this is a very useful concept for ML development flows.

A lot of organizations still use a single VM for training jobs, which although might work at small scale, is not the best way to move ahead. In my opinion, organizations should always use pipeline (DAGs[43]) for defining their project flows, no matter what scale they are at as migration is always painful and it's good to set the correct foundations.

A few major features that using a DAG enables is being able to cache better at multiple stages of a training flow e.g. if my training jobs fails at 1000th epoch, I would like to be able to start my training flow from the last checkpoint saved, instead of having to run my preprocessing for the training data. DAGs also enable a better failure mode for your workflows as, if your main machine that your training (or another important job is running in) dies, you might want to have a good flow that retries/sends custom notification/triggers other jobs, by spinning up new machines.

Another good reason to enable pipeline DAGs is that they enable reproducibility. Let me explain how?

When you are defining a DAG, you'd definitely have to clearly define it in some shape or form. Most of the time this form is a DSL[44] specific (or not) to the pipelining engine you are choosing to run your pipelines with. And then you need a place to store this DSL before sending it to the pipeline engine, so that you have a way to iterate, develop and change your DAG definition as your workflows grow.

The good news is, there's a very well known tool to do exactly this. It has been tested over years for various kinds of applications and is almost synonymous to the process of software development. Yes, I'm talking about git.[45]

One of the easiest ways to implement this is by integrating the pipelining engine of your choice by the git definition of your pipeline DAG with CI jobs. These CI jobs would get triggered every time there is a code change either on the DAG or any of its dependencies in the project repository. The CI job would compile the pipeline DAGs in a stateless environment and upload it to the pipeline engine, further to be utilised by the user.

Now that we've established why using DAGs is a good way to orchestrate our workflows in the ML realm. Let's dive into specifics of which tool would make sense as the pipelining engine.

Kubeflow Pipelines is an open-source tool that we could potentially deploy in one of our existing clusters. Vertex AI is the managed solution by Google using Kubeflow under the hood.

---

[43] [42].

[44] [43].

[45] [44].

Apache Airflow is another open source tool which is designed around orchestrating DAGs for data pipelines and GCP Composer is the managed solution for it.

There are loads of solutions that allow one to take care of pipelines, and to be extremely honest none of them are good enough to completely do the job, but to ensure your organization gets a good solution the works for them, you'd need to carry out a feasibility study to understand what suits the existing infrastructure and future plans w.r.t training, and serving Machine Learning. To provide an example of the evaluation, consider the following text.

In this evaluation, we will be comparing the following three options for the tool of choice for a pipelining engine (Table 2.1).

To summarize the contents of the big table above, Vertex AI offers a fully managed ML platform within GCP, streamlining the end-to-end ML lifecycle. It includes AutoML features, making it user-friendly and accessible, even for non-experts. It seamlessly integrates with various GCP services, facilitating data access, storage, and processing. It also supports end-to-end ML workflows, reducing the complexity of managing different components.

Kubeflow Installation/ Other Open-source in-house installations option although gives us more granular control, but has immense cost of maintenance, and hence are not considered to be viable.

GCP Composer (Airflow) helps manage the operational aspects of running Airflow clusters, there might still be some operational overhead compared to fully managed ML services like Vertex AI. Airflow is not closer and well integrated with other tools that ML needs (like GAR, experiment management, inference services etc.), hence would not be a suitable choice as we would have to make active efforts to integrate airflow with other ML services.

A few issues with choosing a managed service, and in this case vertex ai, would be getting Vendor Lock-in as choosing Vertex AI ties you to GCP. In case we move out of GCP at any point of time, one could still use the same resources we have with a kubeflow installation (Migration is low effort. On the other hand Customization and maintenance is another issue. While being user-friendly, Vertex AI will not offer the same level of customization/visibility and flexibility as open-source alternatives (depths of the k8s cluster that is running the jobs might remain unknown to us) and one would be super dependent on google support in case of issues.

The suggestion in this scenario would be that we should value a streamlined, fully managed ML platform with an emphasis on simplicity and ease of use, Vertex AI would be a strong choice. It provides a comprehensive solution within the Google Cloud Platform (GCP), abstracting away much of the infrastructure management and offering AutoML capabilities for users with varying levels of ML expertise. Vertex AI is well-suited for teams that prioritise a cohesive end-to-end ML workflow without extensive customization requirements.

Table 2.1 Comparison of ML pipeline engines

Feature	Cloud based ML pipelines	Apache airflow	Kubeflow pipelines
Installation method	Managed service	It can be installed on various environments	Deployed on Kubernetes clusters
Ease of installation	Managed service; minimal setup required	Requires setup and configuration	Requires Kubernetes cluster setup and configuration
Integration with GCP	Fully integrated with	Can integrate with GCP services, but not limited to	Can integrate with GCP services; part of Google Cloud Vertex AI
Supported Environments	Cloud-based	It can be installed on various cloud and on-premises environments	Primarily designed for deployment on Kubernetes
Managed Services	Provide managed services for Feature Stores, ML model deployment, serving, and training	–	–
Workflow Orchestration	Provides basic ML workflow capabilities	Excellent support for workflow orchestration	Integrates with Kubeflow Pipelines for ML workflow automation
Extensibility	Limited extensibility beyond the Cloud ecosystem	Highly extensible with a rich set of plugins	Highly extensible, supports custom components and extensions
Community and Support	Cloud support and community	Strong open-source community and Apache Foundation support	Active open-source community with contributions from various organizations
Integration with MLOps	Integrated with Cloud AI Platform for end-to-end MLOps	Can integrate with ML Ops tools and platforms	Integrates with ML Ops tools and supports model versioning and deployment
Monitoring and Logging	Integrated with Cloud Monitoring and Logging	Supports monitoring and logging through various plugins	Can leverage Kubernetes monitoring and logging tools

(continued)

Table 2.1 (continued)

Feature	Cloud based ML pipelines	Apache airflow	Kubeflow pipelines
Cost and Pricing Model	Cloud pricing model; pay-as-you-go	Open-source; self-hosted, potential for cost savings with proper resource management	Open-source; self-hosted, potential for cost savings with proper resource management
Security Features	Security infrastructure and IAM integration	Role-based access control (RBAC), encryption, and authentication	Integrates with Kubernetes security features and supports RBAC
Machine Learning Services	Provides managed services for training and deploying machine learning models	-	Integrates with various machine learning frameworks and tools

Ease of Migration

A very important aspect to be taken into account when carrying out a feasibility study of a tool is to understand what is the cost of migrating from that tool. ML and its operations is a very fragile and changing realm, and its very important to understand what would take to go away from a tool, if it stops making sense for your use case or if your needs change in the future. To understand the migratability, you should consider some factors like the cost of migrating the codebase, integrations, existing assets in the tool, reusability of assets produced/stored by the tool, pain caused to the users of the tool.

Sometimes, you'd have to make do or ignore one of these aspects, but it's good to know the tradeoffs one is making when choosing a tool and going forward with it, as it's one of the biggest decisions you'd make for your of ML Operations.

Migrating from a specific product can take 6 months or more in a production setting, and just switching the tool to a new one and providing related integrations might not be enough, you'd also have to take care of how projects and people are going to migrate to the new tool. This might include, migrating the existing data from the old tool to the new one, migrating backups and providing tutorials and examples for users to move from one tool to another seamlessly.

Reusability

Business needs change every day, especially with the onset of GenAI and other new technologies popping up everyday, its very important to stay at the top of the game. This involves being able to ship new products (in this context Machine Learning projects) as fast as possible without compromising on the quality of the delivered ML system. Not just once, but to be able to consistently deliver such systems no matter the complexity of the system can prove to be the difference between an organization surviving the next wave.

Just like a boxing match or a marathon, where you'd have to prepare for months to be able to deliver at the need of the hour, you'd have to have prepared for a long time to be able to deliver such complex systems back to back. As we can understand that you won't get a long time to change or make a new system, the best one has is to make sure to iterate over small improvements as much as possible, and to gather the collective intelligence and mistakes, it's very important to have common reproducible and reusable aspects to your ML systems/lifecycle/components/abilities.

Let's talk about some of those reusable aspects objectively.

Reusable Project Structure

Having a reusable project structure for all your ML projects can prove to be the most important part of productionising ML. Having a good way to kick start your ML project can shorten the time to production for an ML project from months to weeks.

Maintaining and creating such a template is one of the core responsibilities of an MLOps Engineer. These templates have become common amongst the ML community over the last few years. Tailoring these templates to the needs of the ML systems is another important part of the ML Operations work. These templates also provide a good standardisation for ML code bases as they grow and migrate over the course of a ML project lifecycle. This standardisation is where a lot of infrastructure capabilities and automations tap into (which makes it even faster for ML Engineers to go to production as soon as possible without having to compromise on quality).

To have a quick start, you can just search for an ML project template, get a good variety of flavors, and maybe choose the one that best suits you. Before you ask, *"Well, what factors shall I look for when creating this template? What are the parameters to optimize?"* let me talk about the specifics of this template one by one.

Exploration

As discussed in the EDA section, there's a constant tussle between the need of structure and the need of freedom to explore, and depending on how much you can convince the ML practitioners at your organization, you should try to draw a line. It's not of significant importance to go into the same details of how to implement this (as we have already discussed this), but one has to one way or another figure out where to draw the line of what is EDA, and when is the absolute time to transform your notebooks to proper source code (his is described a bit better in the ML Project Lifecycle section). Having a clear lifecycle stage that defines EDA would help in understanding what is then the sweet spot of drawing that line.

Python Package

As most of the ML work is done in python today (thanks to numpy and scikit-learn) even though rust is catching up pretty fast, the packaging of the code is very important, as it can enable one to unlimited possibilities. And we are gonna talk about python projects specifically here.

Even though it's such a popular language, there are clear drawbacks when it comes to strong typing or sub-dependency management. Afterall, the language was made to provide easy experimentation and not productionising scaled systems. To make this better, there are a number of tools like Poetry,[46] pipenv,[47] conda[48] that have made their way into the usual life of a python programmer.

Having said this, it is very important for an ML project to handle dependencies, have proper lineage tracking and reproducibility in the python code.

One of the best ways to achieve this is to use a single python package per code repository you have. This would make your life easy on a rainy day, as well as make the versioning of ML assets very handy at scale. Using pyproject.toml[49] is one way of making this work, as the whole world is moving away from setup.py as well.

You can find a good example of a simple pyproject.toml that enables one to have a good python structure in the Chap. 3 of this book, under Golden ML template.

Containerisation

Adding on top of the python package layer comes the containerisation layer, everywhere in the world everything is deployed with containerisation (I understand that this statement might be a stretch, but the gist of the statement is that as of now it seems like containerisation is here to stay for a while). From the perspective of operationalising Machine learning, containers provide a good way to encapsulate complexities with the use of pre-built containers. May it be used for creating pipeline components, or storing serving images for prediction servers. Or sometimes used for storing models in an OCI wrapper as well (ORMB[50]).

These containers play a crucial part in ensuring maximum reusability not only at the container image layer, but also at the container cache layer.

---

[46] [45].
[47] [46].
[48] [47].
[49] [48].
[50] [49].

Pipelines

Pipelines affect the ability to reuse resources in a very high impact on the reusability and speed of iteration for the ML experimentation and how fast can MLEs deploy to production.

Having a pipelining engine in place that enables good caching between runs and across projects, can be a good way to ensure that we are optimizing the *'time taken to production'* parameter.

Serving

Replication when in context in serving could be easily misunderstood, with the assumption that:

Should we serve with the same endpoint if the predictions are similar?

Well, NO.

The context of serving here is with respect to reusing as much code/containers/practices when serving a new model out, but when it comes to sharing/re-using infrastructure, its best to keep clear separations between deployments to avoid blast radius.

A good rule of thumb would be to make servers very use-case oriented.

This would make sure that in case something goes wrong, you are only affecting one use case, and are able to keep as many other services live as possible.

My brain correlates this with the the the concept of compartmentalisation[51] in information security. It's just that we are talking about isolation of production systems in this context.

Providing all the boiler plate resources needed to serve makes the infrastructure part of the template complete.

Check out argoapplicationsets[52] (especially git generators) for bonus points of automating serving from these manifests directly.

Manifests for Infrastructure

Most of the times people working with ML or software engineering would not be familiar with the concepts or best practices from the world of infrastructure, and a good way to ensure the quality of services from the conceptions stage is to provide them with a basic set of infrastructure configurations and make sure they can use them to productionize systems.

This ensures that people neither have to bang their head against something they don't understand, nor they have to become pro's at yet another thing and having reusable manifests as a part of the project template provides exactly that.

---

[51] [50].
[52] [51].

Reusable Components

To take the reusability concept to a next level, consider the scenario where you already have multiple projects that are under development in parallel, and all of them start their project workflow from querying some sort of data source, and all the projects write their own logics, and pipeline components.

It would be much more maintainable for the MLEs to use the common fetch_ datasources() pipeline component that possibly the MLOps team has written and maintains in collaboration with the MLEs. This enables common maintenance and implements DRY principle, along with making sure that time is only taken once in writing such a component.

These components could be maintained in a common repository and ML projects could just refer to these components with their URL, when defining that specific project's pipelines. And when the pipelines compiled, the common components get fetched in the workflow of the pipeline DAG.[53]

Reusable Workflows/Pipelines

Similar to the concept of reusable components described above, one could store very common workflows that could be used across projects, and the pipelines could be used as a sub-workflow for project workflows, or these could be conditional workflows that handle the failure modes for project workflows.

In practice, this is a much more mature setup and could be used after you have reached a better stage for reusable components. Having said this, it could also be very well used to measure your MLOps maturity at a relative scale.

Reusable workflows and pipelines can be triggered as a sub-workflow pertaining to certain conditions of execution in the main pipeline. For example, if your batch prediction pipeline did not succeed in execution, you might want to trigger another workflow that tried to schedule backup jobs on a different infrastructure.

Model Selection

This section is going to be pretty short, as this is definitely the skill set of an MLE and they would know what to do best, but there are several aspects of model selection that they might miss and the responsibility of making sure these aspects are considered when developing models goes to consideration when making choices of model selection. Ideally your MLEs should be able to choose any model type or scale and MLOps should provide

---

[53] [52].

enough resources/guidance and support with maintenance, but realistically if there are any restrictions on what can be productionised they should be made clear to the MLEs.

Reproducibility

When it comes to data science models, reproducibility is a critical aspect that should be taken into consideration. Reproducibility ensures that the model can be reproduced consistently and reliably, which is essential for debugging, validation, and auditing purposes. This means that the same model can be generated multiple times with the same input data and parameters, leading to the same results every time. This is particularly important for regulatory compliance, where the ability to reproduce results is often mandated by law.

Also, sometimes it becomes necessary to be able to re-train models and its crucial to get the same results repetitively, and to make sure that you are not stuck in some local minima and retraining the model is not getting similar accuracy or precision that you got once. This also makes sure that there is no potential bug in the code and that the model being trained also that the validation logic is exactly what it should be.

Deployability

When you're building a data science model, it's important to consider not just how accurate it is, but also how deployable it is. In other words, can you put it into production and have it deliver the results you want in a way that's reliable and scalable?

This is where deployability comes in, and it's a critical part of the MLOps process. You don't want to spend weeks or months building a model, only to find out that it's difficult or impossible to deploy effectively. So, you need to think about deployability from the very beginning and make sure that the tools and frameworks you use support it.

You should choose a machine learning framework that can easily integrate with the rest of your tech stack, and preferably use a platform that makes it easy to deploy models as APIs. Also consider tools that allow you to monitor your model's performance in real-time, so you can quickly identify and fix any issues that arise.

All of these factors contribute to the deployability of your model, and they're all things you need to consider as part of your MLOps process. By thinking about deployability from the onset, you can ensure that your data science models are not just accurate, but also practical and effective in the real world.

Ease of Maintenance

More than half of the data scientists and ML practitioners I meet say that model maintenance was the most time-consuming task in the machine learning lifecycle. Most of the time and resources spent on model maintenance were a major bottleneck in their organizations.

By incorporating MLOps practices into the model development process, ML practitioners can ensure that their models are easy to maintain and update. MLOps enables continuous monitoring of the model performance and provides feedback to data scientists, allowing them to quickly identify and fix any issues that arise.

MLOps practices like version control and automated testing can help data scientists ensure that any changes to the model do not introduce new errors or bugs. This helps to maintain the reliability and accuracy of the model over time, which is crucial in real-world applications.

These parameters become more important when a new kind of model is going to be deployed, or there is something special about this specific deployment. In such cases the usual maintenance practices might not suit the deployment and hence could become a maintenance havoc. It is important to consider what would compute an estimate of what is the cost of maintenance for such a deployment. This is essential for ensuring the long-term success of the model and the overall machine learning project.

Model Deployments

Containerization

Containerization is basically like putting your machine learning model, along with everything it needs to run, into a tidy little box (or "container"). Imagine you've got this box that holds the model, the code, and all the necessary libraries, like your own portable mini-environment. You can then take that box and deploy it anywhere that supports containers, and everything just works. No more worrying about weird issues like missing dependencies or "it worked on my laptop but not in production" headaches.

Okay, so let's drop the technical terms for a sec. Think of it like baking a cake. Normally, you'd need to gather your ingredients (flour, eggs, sugar), use the right tools (mixers, oven), and follow the recipe in the kitchen. But imagine if you could just put the entire cake, plus the oven and everything you need to bake it, into a magical box that works anywhere. Whether you're in a home kitchen, a restaurant, or even outdoors, you can open the box, and boom, your cake bakes perfectly every time. That's what containers do, they hold everything your model needs so you can run it smoothly wherever you want.

In practice, if you're using something like Docker (which is one of the popular tools for creating containers), it packages your machine learning model and all its dependencies into this neat "container image." Once packaged, you can deploy it on any infrastructure that supports containers, like cloud platforms (AWS, Google Cloud) or your local servers (Fig. 2.4).

Tools like Kubernetes come into play when you need to handle many of these containers at once. So, if your app gets a ton of traffic, Kubernetes can spin up more containers to handle the load and scale things out automatically. Think of it as having a kitchen assistant who not only bakes the cake but also makes more of them if you suddenly get 100 orders.

What's happening under the hood with tools like Docker is that they use a thing called containerd to manage how containers run and stop, keeping everything efficient. This part makes sure your containerized model behaves the same, whether it's on your laptop, a data center, or the cloud.

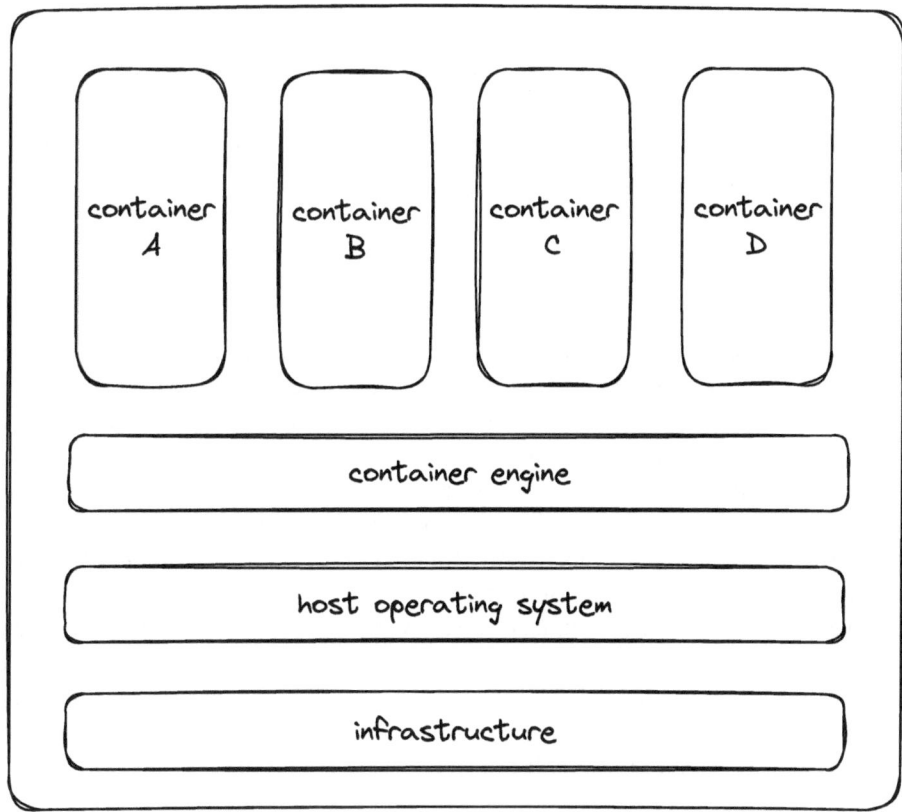

Fig. 2.4 Containerization

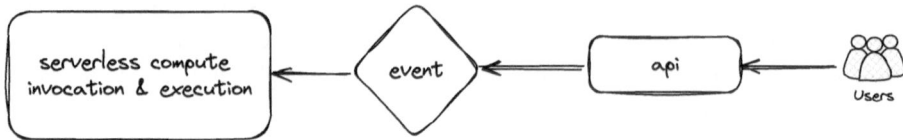

Fig. 2.5 Serverless

There are few other notable tools like Podman[54] and kaniko[55] that serve the same purpose as docker but work a bit differently. Podman allows you to build and run containers locally on your machine. It uses container image build tools like Buildah for creating container images, and can run those containers using runtimes like runC. It's designed to provide Docker-compatible CLI commands, but with more security features, such as rootless containers.

Kaniko builds container images by running inside a container itself. It reads a Docker-file and executes its instructions to build an image layer by layer, pushing the final image to a container registry. Since Kaniko doesn't need a daemon or root access, it can be run securely in a containerized environment like Kubernetes.

In a nutshell, containerization is awesome because it makes deploying machine learning models simple and consistent across any environment. Whether you're scaling models across thousands of servers or just running a small app locally, containers make sure everything behaves the same, every time.

Serverless

Serverless architecture is a cloud computing service where the cloud provider manages the infrastructure and automatically provisions, scales, and terminates the compute resources required to run applications. This allows developers to solely focus on the main source code of the operation that needs to be performed (Fig. 2.5).

One of the cool things about using serverless for ML is cost savings. Since you only pay when your model is running, you're not wasting money on idle infrastructure. Plus, it's really flexible and scales super easily. No need to manually manage servers as your workload increases—serverless just handles it.

Containerization can help with making sure your models work the same everywhere, even in a serverless setup. Think of containers as little boxes that package your model and everything it needs to run. With containers, you don't have to worry about "it works on my machine but not in production" problems. You can take that same container, and deploy it across any infrastrucutre. It just works the same way, consistently.

---

[54] [53].
[55] [54].

So, you can package your machine learning model, along with all its dependencies, into one neat little container. Once packaged, these containers can be easily deployed on a serverless architecture, where they'll run and scale without you having to babysit them.

Serverless lets you focus on your machine learning code while the cloud takes care of all the heavy lifting around infrastructure. Plus, with containerization, you get portability and consistency, so your model runs smoothly anywhere.

Almost every cloud provider has this available as a a service in their offering, and this is really useful for deploying ML models. But this is usually only suitable for small scale models that don't have a huge startup time. This startup time is directly proportional to the size of the container image, and the size of the model.

Microservices

Alright, so let me walk you through microservices architecture in the context of MLOps and why it's pretty cool but also has its challenges.

First off, think of microservices as this way of breaking down a big, complicated system into a bunch of smaller, independent services. Every service does one specific thing and communicates with the others through APIs that have a predefined contract. In terms of MLOps, you can split up the whole machine learning workflow into pieces—like data preprocessing, model prediction, and response post processing; each one as its own microservice, and these services/components talk to each other with predefined contracts (Fig. 2.6).

Let's say your model predictions need more power to process more data; you can scale just that service up without touching the others. This makes everything way more flexible and efficient.

A big win with microservices is that you don't have to scale the entire pipeline when just one part of it needs more resources. So, if your model predictions needs more power because you're dealing with bigger model, you only scale that part. This helps keep costs down and makes managing the whole system easier.

Almost all the tech companies are following the micro-service architecture today.

But, let's not forget the challenges. Implementing microservices for machine learning isn't exactly a walk in the park and needs a lot of infrastructural finnese. It can get pretty complex because you need to have a good grip on things like containerization, infrastructure, APIs, networking across components, etc.. You'll also need tools like Kubernetes to manage the deployment of these services across multiple nodes and tools like Istio,[56] Envoy, (maybe cilium[57] if you wanna go a bit more pro) for service discovery and load balancing. So yeah, there's a learning curve.

---

[56] [55].
[57] [56].

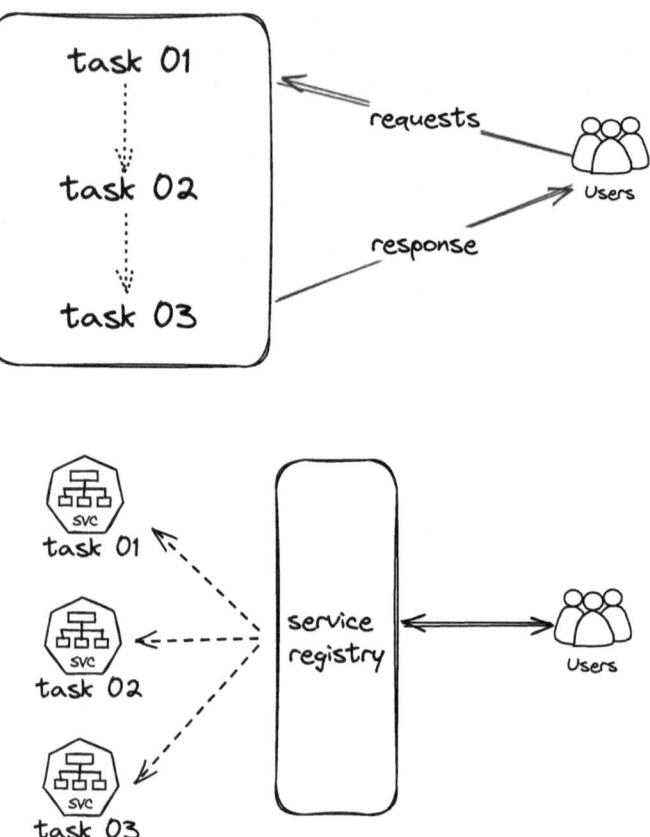

Fig. 2.6 Microservice architecture

You also need to think about whether it's worth the trade-off for your team. A lot of times it might be very difficult to get people onboard an idea that has a significant investment of time and learning tied to it. It requires a certain level of expertise, and managing a bunch of microservices can become a headache if you're not prepared for it.

But when it's done right, the flexibility, scalability, and ability to roll out new updates without breaking the whole system are huge advantages. You can adjust pieces of your pipeline, experiment with different versions of your models, and keep everything modular.

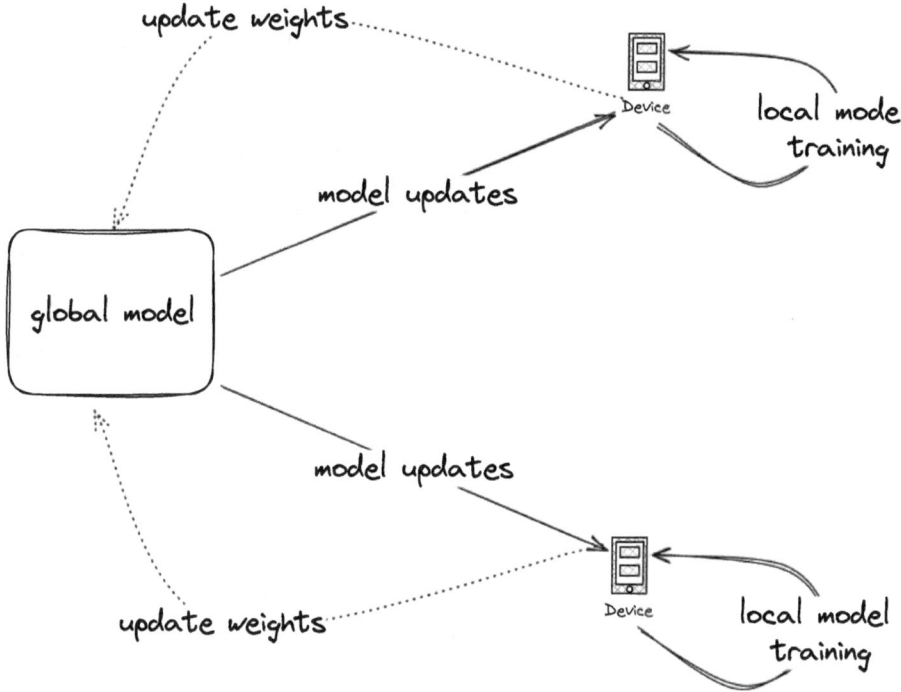

Fig. 2.7 Federated Learning

Federated Learning

Before we begin with this one, I would definitely recommend reading Federated Learning: Building Better Products with on-device Data and Privacy.[58]

Federated learning is a fairly new field, but it has been proven to work for a subset of ML problems. These fields mostly revolve around privacy. If, for some reason you cannot transport data away from the user for training or predictions, you might be interested to look at federated learning. Imagine you've got different data sources to collect data from, maybe smartwatches, phones, or even cars. Usually, all that data would get sent to a central location storage, which would be used to create ML training datasets and then train models. But if it's sensitive stuff, like medical data, or maybe it's just too spread out, or the data is too big to ship around easily (Fig. 2.7).

In federated learning, your local model deployment on the device where the data is collected from would instead use the data to train itself, and then send the weights of the trained model upstream. This eliminates the need of sending the data, as the model weights don't have any Personally Identifiable information, and are very small in size.

---

[58] [57].

As we can see from the federated learning reads from google. They've already started using federated learning for things like improving their on-device experiences like typing predictions or health apps. Instead of sending all your data to their servers, they train part of the model right on your device.

This is very useful for a lot of use cases like Smart Home products we all love to have in our homes. From thermostats, light sensors to security cameras, all these devices generate a lot of data all the time. The other two major industries with these use cases would be healthcare and defense applications. Both these industries would really benefit from privacy of the data generated. Automated cars are also a very good example of this application.

One of the biggest challenges with such deployments and learning is that the devices that data is collected form have a very limited processing power and capability to train models. Also, the applications and development of the training and deployment could needs to be very efficient, robust and lightweight.

There are also challenges around communicating these weights back to the central infrastructure that then aggregates the model weights from the local models to create new global weights which are then send to the local devices.

This is one of the field that is still evolving, but in my opinion this is one of the ML deployments of the future and as things mature. We'd be able to solve these challenges and move towards a more stable deployment and learning strategies for federated learning.

Strategies for Model Deployment

There are different kinds of model deployments in which an ML model can be deployed. The type of deployment is already driven by the consumption pattern for the mode predictions. You could go for:

Batch Deployments

A good number of ML services are deployed in this fashion. This is usually achieved by just using a pipeline that loads the model in memory and then just reads the inputs and predicts in a batch. Usually, such a deployment would write all the outputs in a filesystem somewhere to be picked up by another system of process. These pipelines are usually set to run on a predefined schedule and frequency (Fig. 2.8).

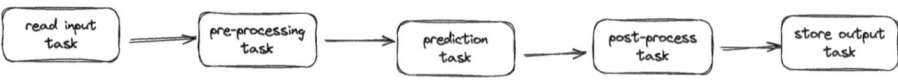

Fig. 2.8 Batch prediction pipeline

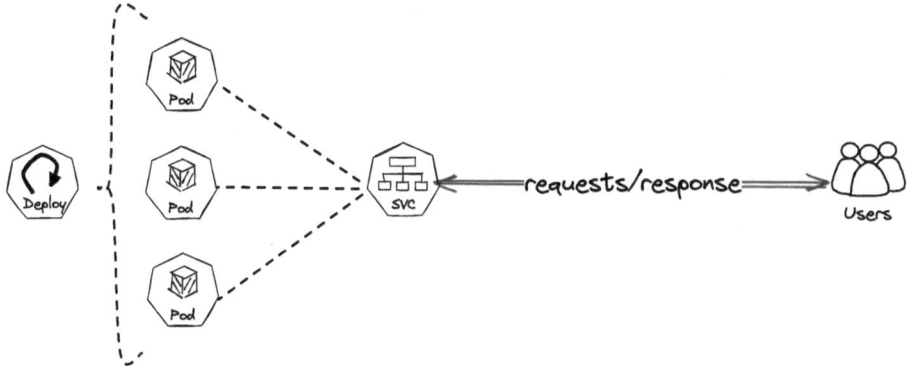

Fig. 2.9 Online deployments

A common reason for using this kind of deployment is when you have resource constraints that don't allow you to keep a service up all the time. Another example could be if your user wants the prediction output faster than your model can predict, but you have a well-known set of inputs. You could possibly use the known set to generate the output, save them in a fast retrieval system like a vector database, and refresh the expected inputs and predictions in the database periodically at a frequency acceptable to the use case.

Online Deployments
Real time or online deployments are the ML deployments where a service is live all the time, waiting for an input and responds with the predictions as a response within a very short time (Fig. 2.9).

This is how almost most of the services work today, with REST or grpc. One drawback with this could be that you need to have enough serving resources available all the time, even if your resources are idle and the model is not actively serving prediction requests.

Serverless Deployments
To address the resource idling issue with online deployments, you can use a serverless deployment as well (as explained before). A lot of cloud providers provide this service where you can define the source code, and the cloud services provide the infrastructure to run it. When a request lands for the server, that's when the source code is invoked and executed. These can be really efficient in saving costs by avoiding idle resources. But these don't work if your model is really large in size and the time taken to load the model in memory is very high.

Embedded Deployments

The previous deployments are the ones where the author of the model has the luxury of deploying it in the cloud. Sometimes, you don't have this luxury due to data privacy or prediction speed reasons. Most such applications are very mission-critical, like ML deployments in space shuttles, networking access points, cars, etc. These kinds of applications need you to deploy models as close to the infrastructure of the end user as possible, hence running the models on embedding systems directly. These kinds of deployments are very dependent on the end device and usually need a lot of awesome engineering work to work at scale (Fig. 2.10).

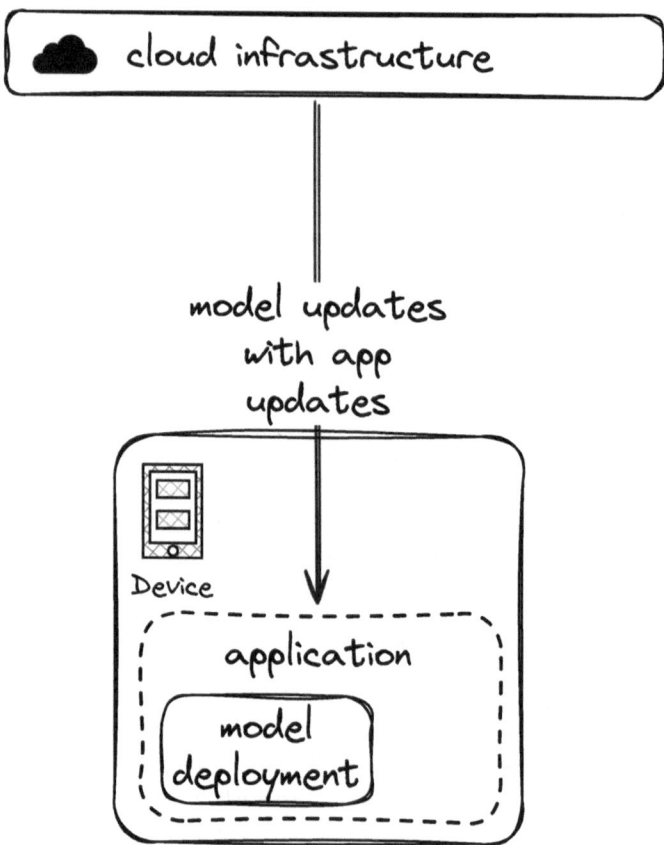

Fig. 2.10 Embedded deployments

Multi-Modal Deployments

These are the kinds of deployment that run multiple deployments at the same time either to enhance the prediction quality using ensembles or to serve complex ML problems. This can be deployed with any of the existing methods discussed above.

Inference Serving Frameworks

One of the core metrics for mlops to optimize is the time taken to production for ML models, and this can be achieved by using as many reusable parts of the puzzle as possible. This gives a lot of head way for MLEs to jump a lot of first steps and leads them to focus only on the aspects of the ML development that is either directly related to the core logic of the business problem they are trying to solve or in researching what new ML core capabilities could be built in the organization (these core capabilities would be used later as reusable backbones behind a lot of specific solutions).

Keeping the recently discussed concept of reusability in mind, when we have to deploy ML models to production, it's really good to not have to think about a lot of common aspects. Most probably your deployment would load a model from somewhere in memory of a storage container/pod/(let's call it a serving unit), and then execute a forward function for the network of a predict() function, and return the response back.

Most of the times this would be a fast api wrapper, the good news is that most of this can be easily done using inference service frameworks like Kserve.[59]

An inference service framework provides a very easy interface of serving an ML model. In the most basic usage of such services, the user needs to simply fill these 3 important parameters:

1. Name of the inference service
2. Path to the model files in an object storage
3. The type of model

The inference service would take care of all the making sure that the model files are fetched from the object storage and necessary k8s resources (like a k8s/service, istio/virutalService, k8s/deployment etc.) are created automatically for you to start serving requests.

This would obviously not suffice all the production usage of such services, and hence some customisation would be needed for different model serving types. One common change you might wish to implement is using a custom image that is used by the framework in the deployment for ML. These custom images might include a specific type of logging or metrics which are tailored to the needs of your organization's central infrastructure.

---

[59] [58].

Testing in ML

Believe it or not, testing has been the most neglected aspect of machine learning. The number of companies I've seen that are operating on the most fragile and untested code is way more than one would expect, to be honest (and I'm saying this from personal experience of knowing organizations I've had the privilege of working closely with). In my opinion, it is the nature of the personnel involved. Most people come from a very academic, research, or number-crunching background, and it takes time for them to realize the importance of set software engineering practices.

One of the common reasons is that ML space develops so fast, and new tech comes alive every second in this world. To be able to beat this, a lot of organizations just delve into the mode of ship it fast for various reasons (sometimes to appease a shareholder, sometimes to prevent it from being obsolete, sometimes for adding a feature before the competitor, etc.). Even though these reasons may seem important enough to ship ASAP, it's important to understand that doing ML wrong can lead to turmoil as well. This can be viewed as an optimization problem as well. Where.

*Speed * Quality = U, Where U is always a constant*

Speed and quality are the forces acting on a see-saw with two buckets of water kept on either side.

Now, depending on the need of the hour, you might have to dial down one specific criterion. But make sure that you dont let the water of your buckets fall down too much. You might want to make sure one side goes a little bit down, but you'd really have to make sure that quickly make the see-saw balance back.

From the perspective of ML Operations, it makes sense to push for the quality of the products as much as possible depending on the needs of the hour for the business. Sometimes, you might have to give up on quality and bite the bullet when things make it to production. But make sure to agree that *'Let's ship it fast'* only when *'as soon as it's shipped, let's start working on quality'* is well understood and agreed upon by all the product stakeholders.

Having said that quality is the first priority, one of the best way to make sure we deliver on qulity is to have the best test in the world available for models we productionise.

Unit

Nothing new here, let's write tests for the single units in our code base, be it a training job, or a model serving code.

This is nothing new for you if you are coming from the world of software engineering, and you might be thinking why I am talking about such a basic thing, but let me tell you, if you are working as an MLOps Engineer, you'd need to reiterate, teach, make people realize the value of and sometimes put your foot down on unit tests, of functions/modules/ etc.

Make sure these tests use a lot of mocking and test data; these tests are meant to test a unit as an individual entity without any external dependencies. Some people would also suggest focusing on making sure that people in your organization stay as close to Test Driven Development as possible.

If you wanna know more about unit tests and wanna know about if they are important, please make sure to go through the answers on stackoverflow for this[60] question.

Integration

Once you have finalized all the unit tests as an independent entity with loads of mocks and patches, the next step in the realm of testing is to test how individual components play out together. These tests should be designed to test how your functions and submodules interact with each other.

Make sure to prioritize testing for edge cases, as optional arguments at loads of times cause unprecedented large issues at the worst times possible (you can read a recent example's root case document here[61]). Such issues (and many more) could be easily caught early when writing smart interaction tests. Essentially, *don't use all your smartness to write code. Save some for tests.*

As already stated before, as an MLOps person in any organization, you'd come across a lot of situations when folks want to move ahead without writing integration tests, but it's your job to make sure people understand the need, take out the time for implementation, and make sure that standards are followed. As the MLOps personnel being the test police is not gonna work or scale, it's best to make sure that the ML/DS community in your organization understands the real depth and meaning of such aspects. That would lead to everyone being the test police, and that's the best way to move ahead with a community that builds robust software together.

Smoke

Smoke testing[62] for ML comes in handy for two major reasons:

- As ML is mostly written in a non-compiled language, it becomes important to make sure that the software is runnable and works.

---

[60] [59].
[61] [60].
[62] [61].

- As deterministic responses and Machine Learning don't go together, its important to understand that the responses a ML system is generating are correct responses with all the possible inputs.

Corner Cases

Best testing can only be carried out for ML when you have tested the system for the extremities. We are not short of examples of ML going wrong, and most of these examples would have never existed if the systems were tested for extremities. Let me describe in detail the levels of extremities I am talking about.

ML System:

Consider a system where one is making a model that looks at the images of a parking lot and tells how many vehicles are parked and what model /brand of vehicle is parked. Let's say this would be used in car dealerships/showrooms/warehouses to have a live inventory management system.

Corner Case:

- Make sure to test that the system can handle getting images of birds/forests/random animals/humans to make sure that in case of an issue in the live feed of the cameras, your model should be able to handle this gracefully, saving this doesn't seem like a vehicle storage facility.
- There should be an easy way to configure the model to filter brands of cars as a default option as a system parameter, which is used live.

ML System:
Consider a system that is an LLM powered chatbot, which makes sure that it can answer about all the plans/offers a company has available for all their customers. As the plans and offers are regional and depend on the location of the user, the correctness of the information given by the bot is very important here.

Corner Case:

- Make sure that when this chatbot is asked about things that are irrelevant to this use case, it doesn't respond to them.
- Make sure that the prompts used to test this chatbot are properly stored and tested automatically when a new rollout of this system is triggered.
- As the answers of this chatbot are specific to the region a user belongs to, it's really important to test this when someone is trying to access this using a VPN or an in-flight user.

ML System:

Consider a system that takes in audio and tells you what the best music genre would be to describe it. Absolute correctness is not of the utmost importance to the model here. However, this is used worldwide by a lot of audio engineers to ensure that they have a good understanding of different music genres from across the world.

Corner Case:

- What format is this audio accepted in the ML system, and can one spoof a completely different input as an input to the system?
- What happens if I send in valid but corrupted audio to the ML system?
- Does the system have a checklist of valid genre names that it has to validate against before sending a result to the user. This would make sure that the model cannot come up with new and potentially problematic genres.

Post Deployment

When we talk about ML systems, having post-deployment tests would be really important.

It could be debated that this correlates with the stage of a test more than the type of the test, and that debate is true to an extent, but making sure you have tests that run once a deployment is up is very crucial with respect to ML models (as mostly ML models are pulled from another storage into the memory of the pod serving the model), to understand that once a model is live, is it gonna behave as it should. To ensure this, its worth while to construct and add some post deployment tests that makes sure that the first few requests that hit the model for prediction are some sane and insane requests. Once these requests go through the deployment properly is the time the server is said to be ready for serving live traffic.

This might be reminding you of a canary or blue green deployments, and it is sort of that but for testing ML predictions of the service.

Pre Re-Training

These tests are supposed to ensure that the expectations we have from our data are being made. Whether it's related to the contract of the data or the distribution, issues with the data contract could be easily brought to light either due to jobs failing somewhere, but when it comes to the changes in distribution of data its really important to understand the new data you are pushing a 'RETRAIN' button with as this could lead you to spending big bucks on an expensive training job which will land you with nothing at the end.

There are few tools that could help you with validation of such data issues and these checks need to be in place alongside the existing data platform checks at the consumer end of the data. Sometimes these tests might be checking the same thing the data platform does at a bigger scale, but it's better to verify the expectations from the data at the initiation of the training job instead of relying on your Data Engineer friends to do it. Also, a lot of times, the specific expectations from your model might not be generic enough that the data platform has thought of them beforehand.

To implement such a check one could use tools like great expectations[63] or evidently.[64] Great Expectations can be used easily by adding it as a prerequisite to your training step in your pipelines, and if the data doesn't confirm to the expectations (your defined contract), it would fail, leading to your training job not starting.

Evidently can be used easily to make sure that you can visualise the state of data and clearly detect drifts and issues with distributions to make sure that your data matches what you think.

Maybe it's a good way to mix both these tools together and make sure that re-training jobs are well fed with the healthiest data that they deserve.

Pre Production

Pre-production tests are the best place to make sure that you are not shipping unexpected surprises to production.

These tests are as simple in intuition as *'Have I tested all the things I need to? Am I confident enough that this would work as expected and won't give me surprises once deployed?'*.

There are multiple ways of implementing such tests, ranging from having a simple Excel sheet that checks off if you have done unit tests, integration tests, and corner case tests to a completely automated solution that makes sure that all the tests that have been done in the previous stages have passed and that we have a final check to say YESS! This should go to production.

An effective way of handling this would be to have a clear production readiness review process in the organization.

---

[63] [21].
[64] [62].

Production Readiness Review

Taking Production Readiness Review[65] from the Google SRE engagement model into consideration as a final check before an application is deemed fit to be sent to production is very effective way of making sure that only and only top notch quality goes to production and gets maintained there.

If you are working at an organization where this seems like a dream, it is the job of an MLOps Engineer to persuade the SREs at the organization to go as close to this model as possible and join hands on this to make sure that ML systems in productions are deployed, maintained, retrained, and archived with the best quality. This is definitely easier said than done, as this not only requires a good tech depth into what the PRR process should look like but also needs a lot of people skills to make sure that infrastructure teams are on the same page and that the product teams would be happy adding one more process to the road to production, and not to forget the developers/ML practitioners that would have to go through this process.

But as it's said, "Great things are worth the wait and the hard work."

Another major blocker in reaching this state would be the length of the PRR process, and the validity of all the questions in your PRR with respect to the system going to production. But the good news is that this is in your and SREs hands to make sure that this process is as smooth and painless as possible for the team owning the system. Also, it's important to iterate on this process and make sure feedback is heard from people.

A simple, sane, and malleable example of the questions to ask in the PRR process for ML systems can be found below. This would need to be modified according to your organization and teams.

- Is this added to the catalog of the organization as a production service?
- Is CI configured for auto deployments?
- Does the CI run unit tests?
- Is the deployment stable in the development environments?
- What hardware is the application running on?
- Are there any specific gotcha's that one should be aware of about the hardware?
- What is the chosen strategy for redeployment? Canary? Blue Green? A/b?
- What other dependencies does this deployment have? Third party and in-house?
- Does the service have metrics, logging, dashboards, and alerts setup according to the organization's norms?
- Are there any single points of failure for this application?
- If this application is down, what is the effect on the organization? Provide a severity score between 0 to 10.
- What is the number of RPS this system can handle? Add links to the load testing documents.

---

[65] [63].

- Where are the runbooks for this application stored? Have the on-call routines and procedures been updated to include details of this system?
- Is this system stateful or stateless? If stateful, please explain why.
- Are there any public endpoints of this system?
- Has penetration testing been done on the system?
- Can model lineage be tracked back to versions of data experiments? Code and artifacts?
- Has the model been trained on complaint data?
- Are there well-defined tests for post-training evaluation and post-deployment tests?
- Have tests been performed with extreme edge cases and unpredictable behaviors?
- What metrics do we use for model deployment and redeployment?
- In case of failures in the ML model, do we fall back to a deterministic version?
- Are you following all the ML test recommendations from MLOps? Elaborate if not.

Maintenance

Imagine you just built a super-complex model that is performing amazingly well in production and real-world test cases, and you are really excited to close off this chapter and move to the next problem you wish to work on...... .

Congratulations! But wait a minute, are you done with the existing solution?

Not really. You see, model building is just one part of the story. The real challenge begins when your model is deployed, and you need to ensure that it continues to perform as expected. That's where model monitoring and maintenance come in. In this phase, you need to keep an eye on your model's performance, identify potential issues, and fix them as soon as possible to avoid any disruptions. It might sound like a daunting task, but with the right tools and strategies, you can ensure that your model runs smoothly in production and continues to provide accurate predictions. So, let's dive in and learn more about model monitoring and maintenance!

Once a model is deployed, it needs to be continuously monitored to ensure that it is performing as expected and delivering accurate results. The monitoring process involves collecting data on the model's performance, identifying any issues, and taking corrective actions to maintain the model's accuracy.

There are several strategies that can be employed to monitor and maintain machine learning models. Some of these include:

1. Monitoring Model Inputs and Outputs: This involves tracking the model's inputs and outputs to ensure that it is behaving as expected. By monitoring the inputs and outputs, data scientists can quickly identify any issues and take corrective actions.

2. Tracking Performance Metrics: This involves tracking key performance metrics such as accuracy, precision, recall, and F1-score. By monitoring these metrics, data scientists can quickly identify any changes in model performance and take corrective actions.

3. Automated Alerting: This involves setting up automated alerts that notify data scientists when the model's performance drops below a certain threshold. This can help data scientists quickly identify issues and take corrective actions.

4. Automated Re-Training: This involves setting up automated re-training pipelines that automatically retrain the model when certain criteria are met. This can help improve the model's accuracy over time.

5. Version Control: This involves version controlling the model and its associated code and data. This can help data scientists quickly roll back to a previous version of the model if issues arise.

6. Data Drift Monitoring: This involves monitoring the distribution of input data to the model to ensure that it has not shifted significantly over time. If significant data drift is detected, the model may need to be retrained on the new data.

7. Human-in-the-Loop Monitoring: This involves having human experts manually review the model's output to identify any errors or issues. This can be particularly useful for models that are deployed in high-risk domains.

These are just a few examples of the strategies that can be employed to monitor and maintain machine learning models. The specific strategies used will depend on the specific requirements of the model and the use case and no matter what and how many strategies you chose from the above lot, you'd have to make sure that you are monitoring and observing the correct parameters.

Metrics

Making sure that you are tracking the correct metrics of the models deployed is the most granular part of the observability stack, and is one of the most important one's in my opinion. If you are not collecting the correct metrics then even an extremely sophisticated and efficient observability stack could end up being of no use.

With that being said, there are some standard conventional metrics that the world of software engineering has laid down for the world to follow, and until now there have not been any standard set of ML metrics. The reason for this stems from the variety of ML applications in the ML world, which is such a fast-paced industry itself.

Let's look at some metrics that would be useful when deploying a Machine Learning Model:

Classical Software Engineering Metrics

These metrics are commonly already tracked in any well maintained piece of software. These metrics allow an insight into the hardware the software is running on, and how healthy is the server running over the course of time.

- CPU usage
- GPU usage
- Number of requests per second
- Number of valid requests
- Rate of 5XX responses
- Rate of 4XX responses
- Avg response time
- ...
- ...
- Others...

Data metrics

As data is one of the most crucial parts of a machine learning system, it's important to make sure that our expectations from data are met every time we try to re-train a model. Check the data section above for more details. Let's look at some metrics that make loads of sense from a Data Platform perspective.

- Rate of outliers in the data distribution
  It is very important to understand the rate of outliers in specific datasets across time. This would ensure that one understands the details of when and how the distribution of data has changed over time.
  In case there is a spike in the value of this metric, it is very important to get an alert, and to propagate the alert to all (verified and potential) data consumers about the issue with the data stream. This alert should be followed by a detailed investigation to understand the cause of the issue and either fix or inform the consumers of a way forward when using this data.
- Percentage of Null values
  The rate of null values in a dataset is very important to point our potential issues with the ingestion steam, or to assess the quality of the data over a period of time.
  This is very useful information that should be provided to ML teams on demand to make sure that they are aware of such issues with the data and are able to make decisions for usage in production or else.
- Rate of failures for data pipelines
  Rate of failures for data pipelines is also another crucial metric that would help the data engineers to make sure that there is nothing fishy happening with a specific pipeline

or the pipelining tool itself. These metrics would be used for creating alerts that are directly targeted towards the teams that are managing the data infrastructure for the organization.

ML Model Training Metrics

Metrics are generally associated with the deployment of a model, but it's very important to make sure that we are paying the same attention to our training jobs as well. In a lot of organizations the pipelining engine used for data and ML training pipelines is the same, and in such scenarios the metrics discussed in the last section could be reused a bit with the addition of some new metrics that are specific to training jobs only.

A good thought experiment here would be to think if the experiment management tools would be best for making sure that training metrics are tracked properly.

There are some training artifacts that need to be tracked and are usually fairly commonly known to be tracked. Most of these are related to the metrics of understanding how well the model is performing. However, there are more metrics that need to be tracked as well, like.

- Training machine metrics
  System metrics for the underlying hardware that is used for training the models. These metrics are generally used to understand the health and capacity of the infrastructure. These metrics should always be tracked to understand when a new machine should be spun up as a backup or to take some preventive action for the training job to continue for longer durations of time.
- Training metrics
  Apart from tracking the training jobs, it's important to make sure that some meta metrics are tracked. These metrics would help us understand how well our training jobs work over a longer period of time. These would also be very important to understand the maturity of handling larger compute jobs.
  - Number of retries for training jobs
  - Rate of warnings in training jobs
  - Rate of errors in training jobs
  - Avg duration of training jobs.

ML Model Deployment Metrics

Once a model has been trained, it needs to be deployed in the world as well and to deploy it scale, one needs to make sure that the model is tracking a lot of metrics related to the specific problem set a machine learning model is trying to solve. Let us take specific examples of types of ML systems that are deployed in production.

- Recommender systems

If one is deploying a recommendation engine that spits out some recommendations with a confidence weight next to it, then the user is free to choose one of those recommendations according to their use case. One should be tracking:

- Number of empty list predictions (when the recommender could not come up with any recommendations that cross the confidence threshold)
- Avg number of recommendations that cross the threshold line (valid recommendations)
- Number of times none of the recommendations are not used by the user at all.

- Feedback loop
  - Rate of downvoted predictions by user
  - The rate of responses that were not downvoted but not used by the user (this could be the user asking the same questions against in the context of a chat-based LLM model)
  - Avg number of wrong recommendations.

Logging

Now that we have discussed the metrics to track let's move a bit towards logging aspects of the code. Logging can look very simple in its principle but can be a very difficult aspect to nail down perfectly in practice. A lot of questions around logging are usually not very simple to answer, and these questions being highly dependent on the context doesn't make it any better. Let's look at some of these questions:

- What should I be logging?
  In short, it is anything that can be logged.
    To elaborate a bit on this, Logging almost everything has its side effects, and storing logs and managing them at scale in infrastructure could drive up costs. But it's the responsibility of the person writing the code to make sure that proper logging configs are set in place, and the default mode of the application is to only log INFO, WARN, and ERROR logging levels, and when there is a need, people can easily switch to logging DEBUG as well and get almost all the logging possible from the application.

- What level should I use for logging?
  Setting the correct level for logging is very important as this can slow down not only your application but also the central logging infrastructure (or increase the cost of storing these logs).
    It's very intuitive to understand what to write under WARN and ERROR logs, but for INFO, it's important to log stages a request has gone through (like pre-processing of vectors completed, prediction initiated etc.)

These stages, when augmented with a proper tracing ID to be able to trace one request throughout your system would make the world a much better place for all the people that handle on-call in your organization.

Logging the correct amount with the correct levels could be a major cost saver for your log storage infrastructure and reduce the load on central infrastructure applications.

- Should I include timestamps in my logs?
  If one is using a structured and well defined logger (preferably used within the organization with as similar structure as possible across teams), it should take care of using a single pre-decided timestamp format.

- Should I include module names in my logs?
  Yes, Including module names in timestamps can make a big difference. Logs are used most of the times when you are trying to investigate something that went/is going/ will go wrong with your application. In such scenarios it's very important to have information about which module (part of the complete application) is this log being emitted from.

- What ML-specific logs must have for a production ready application?
  There are certain aspects that need to be logged for ML systems, such as the dataset name and version used, model name and version used, hyperparameters used, time taken to infer a request, etc.

- Should there be a common convention for logging across the organization?
  Yes, as mentioned above this can be very valuable in case of on-call issues, as the person looking at the logs has their eyes already oriented to look at one specific structure of the logs, which would directly relate to the time taken by them to understand the root cause of the problem.

  This can also enable cross team collaboration, as it becomes easier for people from other parts of the organization to understand what is happening with applications that they do not necessarily maintain.

  There might obviously be certain settings that might only make sense for ML specific applications/teams. This would mean that teams/functions could have their own extended implementation of the base logging module. Something like:

```
from myorg.logging import logger # for normal logging
from myorg.logging.ml import logger # for ML specific applications
```

Could be a good and simple way of implementing the central logging module. Maintenance of the base module should be taken care of the central infrastructure team, whereas the specific extensions could be maintained by the user teams/functions.

Alerting

The first thought that comes to my mind is on-call, *Who should respond if there's an alert after working hours?* I'm gonna leave that question pondering in your brain, while we talk a bit about alerts and how they are different when it comes to ML specific applications.

The best way to make your ML alerting infrastructure more mature would be, do an evaluation of the current state of affairs, and then draft a clear path to maturity.

In order to evaluate the current state, a good way could be to try and deploy a model bigger than the organization has ever deployed. This would not only bring out issues with alerting but would also make sure that you understand what the current infrastructure is able to handle. Also, make sure to add alerts for almost all failures or downtimes on this model deployment. These alerts could be routed directly to the infrastructure/ML/ MLOps team to be solved (make sure the severity of these alerts is low enough that this is not considered as a prod workload). The icing on the cake would be to have another service randomly invoke this model with different kinds of inputs with as many variable parameters as possible.

Setting up such a service (and making sure to note down all the pain points, blockers, and issues) in the first few weeks when you are still getting on board is a really good way to populate your MLOps backlog.

Believe me or not, I've been in situations where I've been asked, "Why should we focus soo much on alerting??". I remember involuntarily coming up with an answer (which I though was very rude, later on), and saying "Well do you drive your car blindfolded?". Having good observability with alerts are those eyes you need to be able to drive and swerve properly.

When it comes to tackling alerts in an organization, it really helps to chunk the alerting stack into multiple chunks. One example of these chunks could be alerting expectations at your organization. Let us take the following example, where we state the expectations from an ML application w.r.t alerting standards. Alerting for ML is classified into the following two stages.

Depending on how many of the following standards your ML application follows, you can estimate the alerting maturity of your ML application.

Alerting Foundations

An ML application is said to be in the Alerting Foundation Stage if the application:

1. exports system metrics such as (these metrics could be available as a standard from the existing infra platform)
2. has a dashboard for visualizing the metrics
3. Has baseline alerts configured for:
   a. Rate of 5XX response codes
   b. Rate of 5XX, per path
   c. Client Errors of 4XX for more than 10%
   d. avg. response time above 1 second for the last 10 min
4. has CD failure alerts for argoCD (or any other CD tool like circles, etc.)
5. Alerts are diverted to either #ml-alerts-dev or #ml-alerts depending on the environments in which the application is deployed.

Alerting Advanced

An ML application is said to be in the Alerting Advanced stage if the application:

1. Has custom metrics exposed for ML applications to understand the following behavior:
   a. The server is not giving good recommendations according to users
   b. Metrics around data drift and prediction drifts
   c. ...
   d. Other relevant metrics, depending on your application.
2. has a clear definition of priority in the alerts w.r.t (all dev alerts could be automatically considered as P3)
   a. P1—Fix immediately
   b. P2—Fix next business-day
   c. P3—Add to the backlog
3. has a clear alert ID specification with the format of < service-name/alert > , e.g., "abc-service/cpu-high."
4. has runbooks for all exposed alerts (this would inform people on-call to understand steps to be taken in case of an alert).

A very important practical consideration to make is the channel that is used for alerting. A lot of times, the number of ML teams you are dealing with might be a huge number, and these teams might come and go out of existence due to re-organizations (these are not new to the corporate world). Hence, making sure alerts are being diverted to a place people are going to look at and respond to can become a challenge, especially if you are not using an incident management tool.

Even in the scenario of an incident management tool being used, it might be necessary for low priority alerts to not go to the on-call team, and to be diverted directly to the team that is maintaining the system in question.

In such cases, diverting alerts to slack channels has been a goodISH hack for such things to work, and making sure that all the ML teams come together and look at a (or two at max) channel should be the job of the MLOps person to set in place. As the alerting maturity model above talks about setting alert IDs, specific teams owning the system could be taken accountable for taking action on the alerts that pop up in the central channels.

These channels, if left for individual teams to maintain, could become too many in number and too difficult to track, hence I would suggest one to go for the central ML alert channels approach, this would boost collaboration between teams as well with the shared responsibility of the alerting channel.

Version Control

Version control are 2 of the very innocent looking words that hide one of the biggest engineering challenges at scale. You must be thinking this is not true *'git (and loads of other tools) has solved it'*. No my friend, I would say that you are naive if you are thinking version control for ML is an easy problem to solve.

Git has solved the issue for code sure, but let's throw in models, structured and unstructured data, graphs, plots, training pipelines, logs, servers, deployments, and prediction requests. The challenge would not just be keeping an eye on the individual versions of these, but also how these versions tie to each other and form a reproducible and always traceable (for compliance, re-trainings, and rollback reasons).

If a data explorer has the ability to modulate and play around with data, create multiple versions of it, has the option to pick one of the versions and take the ML app to production with that version seamlessly, we can say that the ML project is handling its data dependencies pretty well.

And if I may change it a bit to make it fit (pun intended), versioning would be.

If one is looking at an alert for a ML application, and has the ability to understand what version of code, data, model, artefacts, deployment, prediction request, is related to this alert ID, then we can say that the version control for ML is designed pretty well.

In case this deployment's model is retrained, if one can just know the retraining job ID and redeployment version looking at the alert, then the MLOps team at the org deserves a raise.

Now that I have set the golden path for how version control for ML should look like in a utopian scenario, let's talk a bit about how you can go closer towards achieving this.

One of the proven working scenarios is to choose a relative source of truth for the version and then have all the versions relative to that, for example:

Let us assume we choose the code versions (git commit SHA) as the source of truth, in this case and then all the assets we generate or use would be versioned either with the same git commit SHA, or would be versioned relative to the Sha, as the commit Sha being a tag or lined entity to the asset itself.

This is a very common versioning strategy used by a lot of the MLOps platforms that you might have seen around the globe. Where no matter what page you are on, you get a small link to the git commit the current assets is connected to.

Another example of source of truth version, would be to chose the deployment version (let's think semver[66] for better understanding) that this model is going to end up as (if it makes it to production) as the source of truth and all the other versions are named in relation to it.

Another way to look at this versioning problem would be to make sure that all the assets get versioned in their own different systems, but as new assets get generated along the line, the versions of the previous assets get added as tags, or metadata information to the new assets. This would make the metadata for the assets grow in size, further down the lifecycle an asset gets produced.

The problem with all these different kinds of systems is that the assets involved are so different in nature that the tools used with the assets are also varied and have different principles. This makes being able to confidently state that all the versions of my ML assets are properly tracked and linked without having any links pointing to tombstone assets a very confident and bold claim, and this claim keeps becoming bigger the more scalable your system is.

A completely different approach to this problem would be to let all assets be stored in these different systems that can generate a unique ID for the assets while storing them. On the other hand, MLOps can maintain a simple graph database that links all the versions as nodes of the graph against a UUID, and this DB would be used as the single source of truth for understanding all the links between versions. This might sound a very rudimentary and basic way, and maybe potentially be maintenance mayhem, but IMHO, this would not require a lot of maintenance, and if the organization already is using graphDBs (or something similar) as a managed service from an existing provider, then this might even go hand in hand with a lot of processes and tech at the organization.

Maybe there is another approach that suits your environment and situation much better, and you'd have to figure out what suits you in terms of maintenance, the existing ecosystem, and the expertise of the ML practitioners.

No matter what choice you make, you'd have to deal with issues where you might have the version number of the assets, but the asset is missing, or you are not able to locate it. Usually, the last five versions of the asset are very important when it comes to alerts and runtime issues with predictions, but when you are talking about re-training, you might want details of assets that were stored last year and sometimes being able to pull these assets front the cold storage could be a big task. A very commonly known issue

---

[66] [64].

within the ML community is not thinking about re-training models and only focusing on getting the first model out. This leads to a lot of problems that surface in the long term. Focusing on version control and instilling the strictness of version links within the culture and technical implementation is very important to make sure the organization doesn't land in long-term issues.

Drift Detection

Drift detection is one of the very important features of a successful ML Operations system. When we talk about drift with respect to ML, we are obviously speaking about data and model drifts. Data drift is a very commonly known term within the data engineering realm, and you'd find n number of examples and tools about detecting drifts in your data streams. As data is the food for machine learning, it's very important to detect and have the signal of potential issues with your data flagged as early as possible. However, it's also important to understand if the model you are using has drifted from predicting the correct things. On this note, let's dive into these drifts a bit.

Data Drifts

If your organization already has a good setup data infrastructure that lets you know about data drifts as warnings or alerts, or if it lets you configure such drift alerts for the data sources of your choice, then you would be in a very good and happy position. In case this is not your happy and healthy situation, then you might want to consider putting some effort into making this functionality available (preferably as an extension to the existing data engineering tools in practice) to the ML users first, and then maybe you'd consider giving feature as a gift to the core data engineering/platform team (as this is a very crucial and advanced functionality that is useful or other data consumers as well).

If you are implementing this feature in a generic fashion, then you might want to consider a very simple interface (YAMLs seem to be in fashion nowadays), where someone could come to a GitHub repository and say that I want a drift alert on this dataset please, and I'd like to be notified about it through this channel, please. This way would make sure that the functionality is used only for ML folks but is generic enough that a user not familiar with the ML world could utilize it as well for their use case.

Another way to implement drift detection is by monitoring the distribution of the input data over time. This can be done by comparing the statistics of the incoming data to the statistics of the training data. If the incoming data drifts significantly from the training data, it may be necessary to retrain the model on more recent data. Another way to implement drift detection is to monitor the performance metrics of the model, such as accuracy or AUC, over time. If the performance metrics begin to degrade, it may be an indication of drift in the input data or a need for model retraining. There are several tools and frameworks available for implementing drift detection in machine learning models, such as TensorFlow Data Validation,[67] MLflow,[68] etc.

This method and these tools are a bit more closer to the Machine Learning side of things and might not be the best when it comes to generalizing this functionality from the data engineering side of perspective.

Another simpler way to solve this problem would be to use a smaller version of such an implementation using validation tools before you start your training and predictions (as discussed in the Pre Re-training section above). Such tools make sure that one doesn't have to implement a full-fledged solution (that might not be the best investment of time). Instead, work on a lower hanging fruit to make sure we sail through the winds until a larger feature set in the data infrastructure solves the issue.

Three of these ways are very valid and would work to ensure that your MLOps infrastructure moves in a very towards more maturity. Deciding on which would be the best way forward for you, is a very situation and circumstance dependent answer. The factors it depends on are maturity of the data platform, priorities of the data infrastructure team, the need of the feature for ML use cases, etc.

Model Drifts

When it comes to monitoring and maintaining machine learning models in production, it's important to keep track of the model's performance over time. One way to do this is through drift detection, which can help you identify when a model's performance has changed significantly from its initial training data.

Drift detection involves monitoring a model's input and output data over time to see if there are any changes in the data distribution that could impact the model's performance. This can include changes in the data source, changes in the features being used, or changes in the distribution of the data.

To implement drift detection, you'll need to set up a system that collects data from your model's input and output, and then compares it to the data used to train the model. This can be done through a variety of tools and techniques, including data versioning and monitoring tools, and statistical methods like hypothesis testing and change point detection.

There are some open-source tools that let you do this today (like evidently[69]), but being honest, this is still a very unexplored and a bit immature territory when it comes to proven generic tools and systems. It stems from multiple reasons, like (data being super specific to organizations and also that ML, in general, is a very fast moving space in itself). With the lack of a well proven working tool, this is one of the problems that have a good amount of initial time and effort investment needed, but once you have reached

---

[67] [65].
[68] [39].
[69] [62].

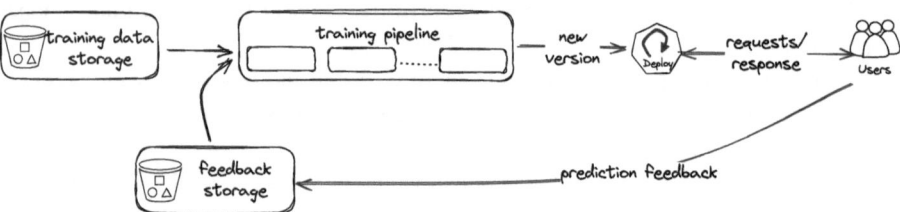

Fig. 2.11 Re-training ML models

a specific threshold of maturity, it becomes a very smooth working feature of your ML system.

Once you've detected drift, the next step is to diagnose the root cause of the issue and take action to correct it. This might involve retraining the model on new data, adjusting the model's parameters, or updating the model's architecture.

Overall, implementing drift detection is a crucial part of maintaining the performance and accuracy of your machine learning models in production. By keeping an eye on your model's performance over time and taking action when necessary, you can ensure that your models continue to deliver value to your organization and customers.

Re-Training

A lot of organizations suffer from the issue of training only once. This is a very bad symptom of a bigger problem they are facing. Making sure your model works once, shipping it to production is not the way to create ML sustainably. The work of an ML practitioner only stops when the model has been decommissioned and archived. Until then, maintenance and monitoring performance is the key to keep delivering business impact (Fig. 2.11).

Naturally, with maintaining the systems and monitoring the models, is followed by the questions of what to do when it starts degrading or performing bad. You iterate!, you iterate and re-train the model to create a new one that is better, or is more inline with the changed needs.

Having this at the back of your mind, and ingrained in the culture of working with ML, leads to people caring about what will happens after the model is deployed in production.

Automated Rollouts

Automated rollouts of newer versions of models are a crucial aspect of model deployment and maintenance. To implement this, there are several strategies that data scientists can use.

One approach is to use a canary release strategy, in which the new model version is deployed to a small subset of users or devices while the rest continue to use the old version. This allows for easy monitoring of the new version's performance and the detection of any issues or discrepancies.

Another strategy is to use a blue-green deployment approach, where the new version is deployed alongside the old version, but not yet serving any traffic. Once the new version is fully deployed and tested, traffic can be switched over to it, with the option to quickly switch back to the old version if any issues arise. ArgoCD now supports such deployments, and making sure that the infrastructure team at your organization understands the need for this feature not just for ML but for the rest of the tech could be a really good way to let them prioritize and deliver this feature for you, and you could piggyback on the existing work that has been done.

In addition to these strategies, automation tools such as Kubernetes or Docker Swarm can be used to simplify the deployment and management of new model versions. These tools can enable automatic scaling, load balancing, and rolling updates, ensuring that the new version is efficiently and seamlessly integrated into the overall infrastructure. As you might be able to notice, almost all of these features are not specific to ML but should be made generally available for all deployments. Making sure that you collaborate with the infrastructure team to make sure these features make it to the finish line can make the life of an MLOps Engineer very happy and healthy.

There exist a lot of tools that enable one to deploy and rollout resources with these strategies, and as these strategies are not native to ML, I would say that most of these are battle tested methods of rolling out newer versions of deployments.

Talking about frameworks not being native to ML, if we look at the other side of the spectrum there exist multiple tools like Tf serving[70] which has an interesting concept of Servables,[71] which acts like an entity that can serve a request. This when combined with the version policy[72] make sure that multiple servable versions can be live at the same time, which makes is easy for a request to choose a specific version of the servable.

Another interesting framework to discuss here would be kserve[73] (formerly called kfserving), which makes Kubernetes based deployments easy but concepts such as model deployments and model mesh.[74] Kserve model deployments make sure that k8s users have a CRD to interact with, which makes sure that k8s.deployments, k8s.services, and related resources are created on your k8s cluster, with the use of a simple YAML that defines your model location, type of model, and some scalability related configurations. This also makes it easy to implement a queuing mechanism for workers, which is needed for models with higher latency of prediction requests with the help of inference batching.

---

[70] [66].

[71] [67].

[72] [68].

[73] [58].

[74] [69].

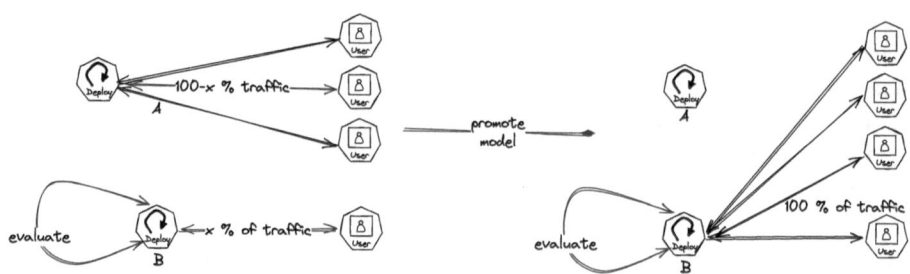

Fig. 2.12 A/B Testing

Kserve model mesh plays around the idea of loading and unloading models according to the need of the requests, and this makes trade offs between the latency of responses vs the resources used under the model deployments. Kserve provides the ability to deploy canary deployments, and one can use them for all model rollouts just like one would deploy any other CRD to a Kubernetes cluster (using infrastructure as code[75] and controlling all manifests that go to cluster with a git branch would be my personal suggestion, and I'm sure your organization's infrastructure team would surely agree with me).

Most of the places use ArgoCD today for their deployments (of fluxCD), and both of those tools provide the option to deploy canary and Red/Green strategies today (either out of the box or using an extension). Maybe the best way for your organization would be to use the core kubernetes deployments and utilise the CD mechanism to rollout ML servers.

Just like all the other choices, we've discussed in this book, it's really up to the priority, use-cases, priorities, maintenance efforts and existing infrastructure to decide what suits as the best way forward.

A/B Testing

A/B testing is a common technique used to evaluate the effectiveness of different versions of a model. The basic idea is to split traffic between two versions of the model (e.g. version A and version B), and measure the performance of each version.

You can use any pipelining engine to implement such a pipeline (Fig. 2.12).

Implementing A/B testing requires careful planning and execution. It's important to define clear success criteria, choose appropriate metrics to measure performance, and carefully monitor the experiment to ensure that it's generating meaningful results. Once the experiment is complete, the results should be analyzed to determine the best course of action. If one version of the model outperforms the other, it may be necessary to roll out

---

[75] [70].

the new version to production, while if the results are inconclusive, further testing may be required.

Being able to deploy a model with A/B tests and being able to deploy all models at scale can be two completely different games to play and obviously until absolutely necessary one should go for a solution where every ML practitioner in the organization can use this deployment/testing strategy without knowing the details of the infrastructure at all.

Such an efficient workflow can only be achieved when there's a framework in place that expects its users to define what model needs to be deployed, what evaluation metric to use, what time scale to evaluate the metric for, and how much traffic to shift to the new deployment. Once these four questions have been answered by the user, the framework should deploy a new version, shift traffic, evaluate the metric, and automatically move the traffic to the winning model deployment, leaving a notification to a channel about the results. In case there is some anonymity or confusion about which model is the winning model, the user should be notified that human interference is needed to make a decision. A few of the known tools provide such functionalities today.

A/B tests have been used in software development quite a lot now, especially in the UX/UI aspects of services. They have depended on A/B to understand what works best for the users and lets them stay more on a service/application. The reason for this being an important aspect of UX rollouts is that there are some real world human aspects of UX that cannot be computed with a 100% surety all the time. Hence such tests help in validating the understanding of user behavior and assumptions.

Similar to the UX world where some unknowns need to be validated with A/B tests before the complete traffic, ML needs to validate some assumptions about live traffic and real data as well. No matter how good the model performs on the test set, and no matter how close the test set is to the real world use cases and data. The real performance evaluation of a model can only be done once it's live, and making sure all your models get deployed with A/B rollouts as a default behavior at your organization can be a good default.

If we go a bit more detailed into these tests, the strategy used for splitting traffic can really impact your tests results. Hence it's important for the ML practitioner to understand how to split traffic, for their model and use case. A few well known methods include time based splitting, geolocation based splitting, random splitting, etc. In essence one has to choose a split that creates a sample representative of the complete population. As in case our sample would not be representative of the population, the decisions made for choosing the winning model would be false, as the model would work better for the part of the total population that the sample represented. Taking a final call on which split is very dependent on the requests and kind of requests a model gets, hence its very important to make sure the requests coming to a model are logged, evaluated and understood as a batch process alongside the maintenance of the model servers.

How to Propagate Bugs Back to the Owner and Propagate Fixes to Prod?

We've already gone over the essentials of MLOps like versioning, data prep, deploying models, and monitoring. Let's talk about something just as important but often ignored, What to do when things go wrong. Yes, I'm talking about bugs and issues popping up in production.

In a solid, production-ready ML setup, you need a clear process for dealing with bugs when they show up. First, when a bug is discovered, you have to figure out who's responsible for that code and let them know fast. There are pretty awesome tools available today at catching errors in real time and showing you exactly where the problem is. So, instead of manually hunting for that one error buried deep in logs, these tools just point you right to the issue.

Once you've tracked down the bug, the next step is fixing it. But don't rush into deploying the fix to production just yet! Tools like pytest[76] or unittest[77] help you write tests that check if everything still works the way it should. It's kinda like giving your code a test drive before hitting the highway—you wouldn't want a breakdown in the middle of nowhere, right?

After that, you're ready to deploy the fix, but not directly to production (we're not that reckless!). First, you roll it out in a staging environment. Staging is like a copy of your production environment, where you can safely test things in a space that behaves just like production. It's your last chance to catch any remaining issues before going live.

When everything looks good in staging, it's time to deploy to production. A rolling deployment strategy is often the way to go here. Instead of updating all servers at once, you update them in a rollout fashion. If something goes wrong, it only affects a small part of the system, and users.

Ideally, you should be using a Continuous Deployment (CD) tool that would automate this process for you. If any issues pop up again, you can catch them early.

And finally, you always need a rollback plan. Even with the best testing, things can still go wrong, so you should always have a strategy to revert back to the previous working version if needed. Better safe than sorry!

Having a well-oiled process for catching, fixing, and deploying bug fixes quickly is key to keeping your machine learning systems stable and reliable. The more automated this process is, the faster you can respond when things go wrong, and that's a huge win.

---

[76] [71].
[77] [72].

Human in the Loop

How Do You Talk to Data Engineering Teams?

Make sure that there are close connections and ties between the data, MLOps, infra, and ML teams. These teams need to work together to achieve long-term success.

This is such a key relationship because, as we always say, the quality of your machine learning is only as good as the data that feeds it. But in reality, that means you need a lot of coordination between your teams to make sure the data is accessible, clean, and in the right shape for ML use cases. It's more than just "Here's the data, good luck!" It's about building that ongoing partnership.

When you're talking to data engineering, it's important not to just throw a bunch of requirements at them. Instead, sit down and walk them through what you need, not just from a technical standpoint but from the broader perspective of why certain data is critical to the ML models you're building. They'll likely care about how clean and optimized the data pipelines are, and you can relate that back to how well your models will perform. Think of it as a regular conversation where both sides are constantly adjusting and refining how they work together.

The other part of this is consistency. For ML models, the quality and format of the data have to be predictable and reliable. This is where the data engineering team becomes crucial because they can ensure that the data pipelines are delivering this consistently, so your models aren't constantly retraining on bad or incomplete data. And don't forget, scaling up is something that has to be front and center—what works for small experiments won't work once you're handling larger volumes or need to retrain models regularly.

How Do You Collaborate with Multiple Teams?

Collaborating with multiple teams would be one of the biggest social challenges in working with MLOps, as you need to collaborate on so many teams and tech aspects. One of the things that has worked for me is always making sure that you are in regular contact with other teams that are a part of the MLOps equation. This communication is with respect to the priorities, future plans, etc., of both teams. This can help a lot in a lot of situations, as just knowing what is happening in the other teams can make you plan alignments and work better.

One other very practical tip would be to just make sure to grab a coffee casually with people as well. A lot of collaborations are based on casual connections people have rather than all the serious business conversations all the time. A coffee conversation goes a long way. Believe me.

What Expertise Should an MLOps Team Have In-House?

Well it's really important to have at least one person that has Machine Learning experience. Having a clear understanding of how ML works and what are the pain points that a ML practitioner has to go through is very important to be able to build, help and aid the development of ML models.

Another important expertise to have in the team, is to have experience of scaling production systems that are resource heavy, and complex. This would be really handy for engineering solutions that work at scale. Also, handling infrastructure when stuff goes down in production is very important to be able to debug in production and fix issues when live model are down with a clock ticking against it.

The third most important expertise to have in house is data-engineering. I am guessing this is not a surprise, as this expertise would really help in setting up scaled and illustrate pipelines that would work for training and batch predictions as there's a lot of common synergies between both those worlds.

So, in essence, a perfect MLOps team would have a bit of data Engineering + a bit of Site Reliability Engineering + a bit of Machine Learning Engineer to have a good setup for success.

Where in the Organization Should an MLOps Team Sit?

Ideally the best place to be for an MLOps team would be to be with the core-infra team. MLOps's roadmap and work should be heavily influenced by the needs of ML, at the same time making sure that the MLOps stays inline with SRE on core-infrastructure roadmap and future.

There are major core infrastructure components (like GPUs, CI workflows for ML, k8s clusters, etc.) that MLOps should aid the infrastructure team in maintaining. There are a lot of other aspects that both the functions should collaborate on as quite a few MLOps functionalities are built on top of the core infrastructure of the organization. Just to understand the length of similarity in work, below is a list of services that overlaps with the core infrastructure team.

- Deployment infrastructure
- A/B Deployments and Rollouts
- Integration with existing backend services
- Experimentation Infrastructure
- Pipelining Infrastructure
- Testing Strategies for ML
- Asset management for ML

Close collaboration is suggested with SRE in order to ensure that MLOps introduces components that fit in the overall technical stack, or there is a huge risk of creating siloed infrastructure that is unmaintainable over time. Whereas there are only a few other collaboration points for data engineering teams (data ingestion, data versioning, feature stores, etc.)

Another important aspect is that as the world becomes more ML-centric, it can be really useful for SRE engineers to get close to MLOps.

Disaster Recovery

Having a clear disaster recovery procedure and systems that automate disaster recovery is very important. Not everything in the disaster recovery setup needs to be automated; it just needs to be in place, and everyone who works with such systems needs to be aware of what to do when a disaster happens. There needs to be a clear document saying how and which systems need to be brought back, and this document should be available to the on-call team all the time. There needs to be a special section in this document called 'Machine Learning' detailing what is worth bringing back with what timeline into consideration. Let's look into this disaster recovery a bit closely from the ML perspective.

What to Recover?

One important aspect to understand is what needs to be recovered and in which order. Just to be clear we are talking about a disaster and not an incident here, so we assume that everything is gone. In such a case, the first thing to recover would be definitely all the model files stored as that's the gold we've been minting with all the hard work.

Once the models have been recovered, the second asset to bring back would be the model servers that would use the recovered models to start serving again as soon as possible. This would at least give us some breathing room. If your live services are serving from some pre-predicted outputs, making sure to bring back those output files would be the top priority so that you can resume operations as soon as possible.

The next asset would be all the other ML assets that you had stored. Once those are back in place, the pipelines would be the next one, as these pipelines might be looking at the previously stored ML asset for various operations. At the end would be the datasets that you've been using to train the models.

How to Recover?

Okay, fine, we recover these resources but how do we do it? There might be limitations to what one can do to recover from disasters, as the dependence of the modern world on cloud providers allows us to not think about these aspects a lot, and rightfully so, as it's very rare for cloud providers to fail with such big disasters. You can still take some proactive actions to make sure you have a better chance at the time of such a situation.

Most of the models would be stored in cloud storage services, and making sure the buckets holding these models are versioned and have automated backups configured, preferably in regions other than the source bucket. This also holds true for the ML artifacts, as you might have artifacts in a self-deployed experiment management tool, but making sure that the content of the experiment management tool is backed up is really important.

When we talk about deployments that are live, it is very important to make sure all these are defined in code and synced from some git-based repository. Another thing to ensure is that all ML servers are completely stateless at all times. Once you have the models backed up with storage, code, and deployment definitions in Git, the whole disaster recovery can be automated easily.

Pipelines are also ML artifacts in a way, and can be easily backed up in the same way. Another step that should be in place is having pipelines defined in your source code and git controlled.

One thought you might have had is, why not bake the models directly in the container image with the server's source code? That is a very valid question. To be honest, until your models are small in size, this can work out pretty sweetly. The issue is in scaling this approach. As the model size grows, the time taken to push and pull the images grows exponentially and is more than just pulling the model from a cloud storage. Another issue is the cost; even though container image storages are usually backed by cloud storage, they are more expensive to use.

Failover Clusters?

Another very important aspect to consider is having failover clusters ready. These clusters essentially would have the same deployments as the live cluster but scaled down a bit to save costs. And incase of a disaster, the first action taken is to just switch over to the failover cluster so that there is minimum disturbance to the users (Fig. 2.13).

If you have an existing staging cluster that is maintained and used with the same piousness as production, you could use that as the failover cluster as well. Ideally, such failover clusters should be in a different zone/region or infrastructure to make sure that we minimize the chances of both clusters being hit with the same disaster.

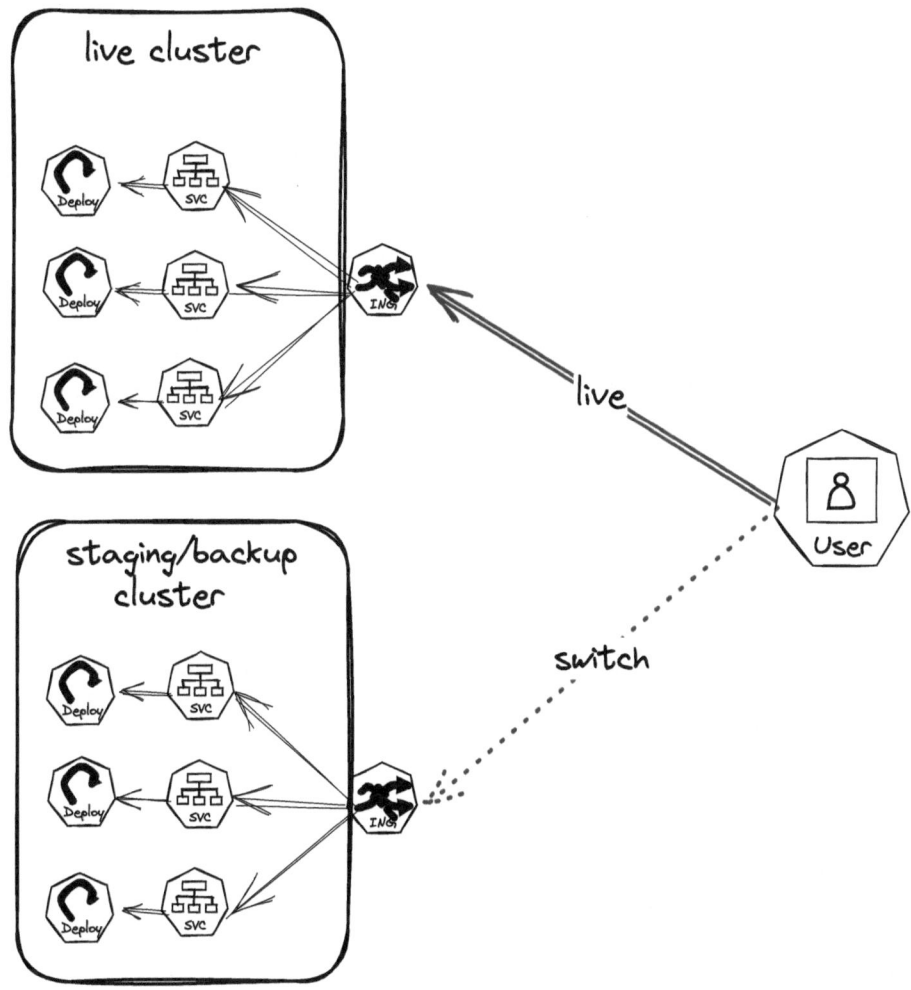

Fig. 2.13 Failover clusters

How Do We Package Big Changes or Releases and Force People to Migrate?

One does not have to work towards the absolute best, but having a positive delta and moving towards the best-case scenario in a way that doesn't cause too much trouble for the existing work MLEs are busy with, is what counts. Introducing large changes in an organization can be really difficult, especially when people are very used to the existing ways of doing things. This can make any change painful for the driver and the receiver, leading to friction between teams and people. Having even a few such frictional episodes can lead to the general feeling of avoidance and anger when new changes are introduced from central teams.

Although there are a lot of human aspects that can be improved on to make such changes better for all the parties involved, that revolve around clarity of communication and detailed planning, but we are going to discuss about what can one do at the technical level to alleviate pains for all the people that'd go through the change and would have to adjust according to new ways of doing their work.

Technically speaking, there's a sweet way of packaging such bigger changes. If you follow this packaging strategy for rolling out changes, not only would developers be eager to adapt to the changes, but they also would be happy with the progress they are able to make with their work. This packaging strategy makes them understand the benefits of the changes by hitting the developers at the most painful part of their existing work. According to this strategy, an ideal change or major release should have.

The Bait

The bait is the change that everyone has been asking for. This is the request you've been getting from the developers for some time now. This is the change of a major release that affects the life of the ML practitioners the most. This has to be something that'll make sure any ML practitioner agrees to make changes in their existing code/ways of working to be able to get this functionality. This is usually the change you've received as the most voted on in your feedback interviews.

The Painful Chore

The other important change is the one that is very painful and tiresome for the users to adapt to, but the core teams (in our case, the MLOps team) have been trying to introduce it. This is the chore that, if introduced independently, would cause a lot of chaos and backlash from the developers, as this is one of the most difficult changes to adapt to, and probably, the benefits of such a change are not directly visible to the developers or doesn't affect the developer experience directly.

Why Should I Align with the Rest of the Org?

This is the change that, if released independently, can get the above-stated question. This might not have a direct impact on the developer experience or a clear benefit to the users or the releasing team. However, if viewed from the aspect of standardization, this change would seem very important as the existing ways/code lacks uniformity in execution. Such changes make the life of the core teams easy, as they can easily assume a standard that is followed across the organization and make new features on top of those assumptions.

The Future is Bright

The last part of the release has to be the future is bright change. This is a change that doesn't solve an issue today but is a foundational step in the direction of solving something later. Most of the time this should be the foundational step towards the bait of the next release.

References

1. https://en.wikipedia.org/wiki/Diffusion_model
2. https://en.wikipedia.org/wiki/Garbage_in,_garbage_out
3. https://en.wikipedia.org/wiki/Cloud-native_computing
4. https://github.com/apache/spark/
5. https://github.com/apache/spark/sql/
6. https://github.com/apache/airflow
7. https://github.com/dbt-labs/dbt-core
8. https://en.wikipedia.org/wiki/Scrat#Acorns
9. https://github.com/apache/parquet-format/
10. https://github.com/iterative/dvc
11. https://github.com/treeverse/lakefs/
12. https://github.com/pachyderm/pachyderm/
13. https://github.com/treeverse/lakeFS/tree/master
14. https://github.com/treeverse/lakeFS/blob/ae44e05c1a61bf3a6931746415001e68dc05a993/docs/integrations/git.md
15. https://github.com/treeverse/lakeFS/issues/2073%23issuecomment-956534835
16. https://github.com/treeverse/lakeFS/issues/2739
17. https://gdpr.eu/right-to-be-forgotten/
18. https://github.com/treeverse/lakeFS/blob/master/design/accepted/gc_plus/uncommitted-gc.md
19. https://github.com/duckdb/duckdb
20. https://github.com/treeverse/lakeFS/blob/3d0db3081411d1afb54fd5317f51a1dd6f44180e/docs/integrations/duckdb.md
21. https://github.com/great-expectations/great_expectations
22. https://github.com/alteryx/featuretools
23. https://cloud.google.com/automl-tables/docs/quickstart
24. https://github.com/tensorflow/tfx
25. https://github.com/pycaret/pycaret
26. https://en.wikipedia.org/wiki/KISS_principle
27. https://github.com/feast-dev/feast/
28. https://github.com/tecton-ai
29. https://colab.research.google.com/
30. https://www.kubeflow.org/docs/components/notebooks/

31. https://github.com/kubernetes/kubernetes
32. https://github.com/mwouts/jupytext
33. https://github.com/kubeflow/kubeflow/tree/master/components/notebook-controller
34. https://github.com/kubeflow/kubeflow/tree/master/components/example-notebook-servers
35. https://github.com/coder/code-server
36. https://github.com/ipython/ipykernel
37. https://www.kubeflow.org/
38. https://github.com/apache/airflow/
39. https://github.com/mlflow/mlflow
40. https://cloud.google.com/dataflow/docs/
41. https://github.com/argoproj/argo-workflows
42. https://en.wikipedia.org//wiki/Directed_acyclic_graph
43. https://en.wikipedia.org/wiki/Domain-specific_language
44. https://github.com/git/git
45. https://github.com/python-poetry/poetry
46. https://github.com/pypa/pipenv
47. https://github.com/conda/conda
48. https://packaging.python.org/en/latest/guides/writing-pyproject-toml/
49. https://github.com/kleveross/ormb
50. https://en.wikipedia.org/wiki/Compartmentalization_(information_security)
51. https://argo-cd.readthedocs.io/en/stable/user-guide/application-set/
52. https://en.wikipedia.org/wiki/Directed_acyclic_graph
53. https://github.com/containers/podman
54. https://github.com/GoogleContainerTools/kaniko
55. https://github.com/istio/istio
56. https://github.com/cilium/cilium
57. https://federated.withgoogle.com/
58. https://github.com/kserve/kserve
59. https://stackoverflow.com/questions/67299/is-unit-testing-worth-the-effort
60. https://www.crowdstrike.com/wp-content/uploads/2024/08/Channel-File-291-Incident-Root-Cause-Analysis-08.06.2024.pdf
61. https://en.wikipedia.org/wiki/Smoke_testing_(software)
62. https://github.com/evidentlyai/evidently
63. https://sre.google/sre-book/evolving-sre-engagement-model/
64. https://github.com/semver/semver
65. https://github.com/tensorflow/data-validation
66. https://www.tensorflow.org/tfx/serving/architecture
67. https://www.tensorflow.org/tfx/serving/architecture#servables
68. https://www.tensorflow.org/tfx/serving/architecture#version_policy
69. https://kserve.github.io/website/0.13/modelserving/mms/multi-model-serving/
70. https://en.wikipedia.org/wiki/Infrastructure_as_code
71. https://github.com/pytest-dev/pytest
72. https://docs.python.org/3/library/unittest.html

The Gold Standard MLOps

3

Introduction

If a data explorer has the ability to modulate and play around with data, create multiple versions of it, has the option to pick one of the versions and take the ML app to production with that version seamlessly, we can say that the ML project is handling its data dependencies pretty well.

In this chapter, we are going to discuss an almost utopian scenario of what ML Operations at an organization should look like. This would be an almost perfect system that might be difficult to achieve but can serve as the north star for any Machine Learning Operations team to chase. Some might also think that the system described in this chapter is a bit too unrealistic to achieve in a practical sense, but that is what this chapter serves to be 'An unrealistic almost perfect system, that one could always take as a roadmap or direction to move ahead'.

In my opinion, a perfect MLOps system would constitute of the following parts:

- A perfect template to get a head start
- A very efficient and cache enabled CI system that can handle mega sized builds
- A seamless CD system
- A clearly defined process for the ML project lifecycle
- A well defined git repo system
- A very efficient experiment management system
- A notebook/report archival system
- A scalable artifact storage
- A scalable vector storage and retrieval system
- An easy interface to consume data for small-scale experiments and large training jobs
- Scaled infrastructure that can handle training and serving jobs of varying size

© The Author(s), under exclusive license to Springer Nature Switzerland AG 2025
P. Mishra, *A Guide to Implementing MLOps*, Synthesis Lectures on Engineering,
Science, and Technology, https://doi.org/10.1007/978-3-031-82010-6_3

- A super awesome MLE/Ops team that has expertise in troubleshooting and system design
- A very pessimistic Red Team.

If we dive deeper into all of these parts of the system, we'll be able to understand the complexities involved in the process.

Golden ML Template

Being able to spin up a new project and have a deployment out in 15 min is still a dream for a lot of developers in the real world today. This is something almost all of the core infrastructure teams strive to provide. The deployment doesn't have to be a full-fledged product. It could just be a simple hello world machine serving responses to a default endpoint. This enables the developers not to have to worry over setting up stuff to take their project to production, and they can easily iterate on the existing hello world application to make it serve the use case they wish to build.

An infrastructure team can have a few templates depending on the use case/language of the project, which could help set up almost all of the infra resources needed to take it to the finish line. Then a dev could just fill in the business logic like a neat lego piece in the project, and make sure it goes to deployment.

There is a lot of a project that needs to be set up at the initial time, from precommits to logging and tracing. There's a wide spectrum of things that a dev has to do again and again in every project, and all of those aspects should be just thrown in a template for people to use. These templates also act as a very good way of standardizing workloads and designs of applications that get developed in an organization. With the standardization, a lot of assumptions can be made while developing the core infrastructure, e.g., one can assume that all the infrastructure manifests of a project will have a label called `app.owner:team-a`, and then the central cloud cost aggregator can easily group resource costs on the label app.owner.

Well, you may ask, we don't start a new project or repository that often, and that is a valid case of when you should not have such a template, but from the perspective of an SRE, having such a template would still be very useful. The good news is that ML projects have a completely suitable lifespan for such a template, and the world has already realized this. You can find a lot of ML templates on GitHub today, and I'm gonna explain what is the golden and probably unrealistic template.–

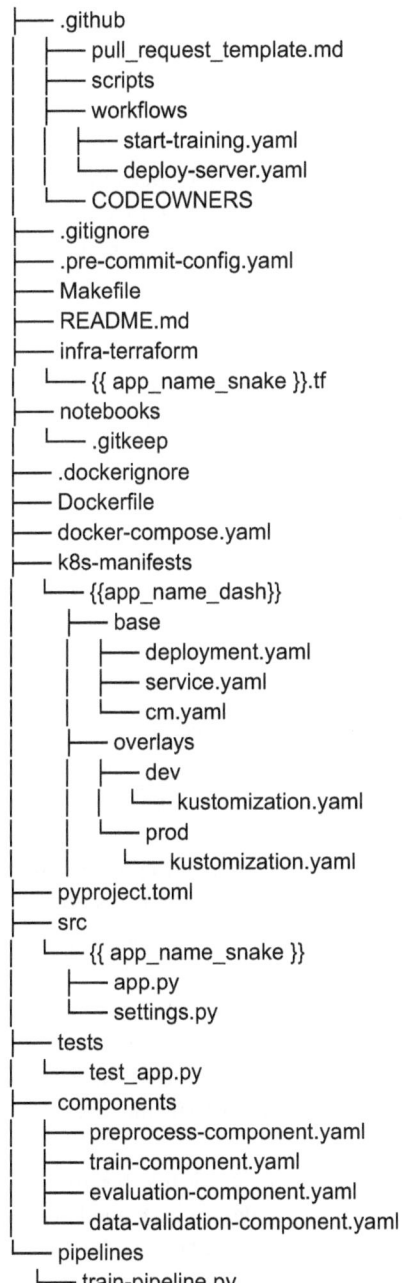

```
├── .github
│   ├── pull_request_template.md
│   ├── scripts
│   ├── workflows
│   │   ├── start-training.yaml
│   │   └── deploy-server.yaml
│   └── CODEOWNERS
├── .gitignore
├── .pre-commit-config.yaml
├── Makefile
├── README.md
├── infra-terraform
│   └── {{ app_name_snake }}.tf
├── notebooks
│   └── .gitkeep
├── .dockerignore
├── Dockerfile
├── docker-compose.yaml
├── k8s-manifests
│   └── {{app_name_dash}}
│       ├── base
│       │   ├── deployment.yaml
│       │   ├── service.yaml
│       │   └── cm.yaml
│       ├── overlays
│       │   ├── dev
│       │   │   └── kustomization.yaml
│       │   └── prod
│       │       └── kustomization.yaml
├── pyproject.toml
├── src
│   └── {{ app_name_snake }}
│       ├── app.py
│       └── settings.py
├── tests
│   └── test_app.py
├── components
│   ├── preprocess-component.yaml
│   ├── train-component.yaml
│   ├── evaluation-component.yaml
│   └── data-validation-component.yaml
└── pipelines
    └── train-pipeline.py
```

src directory

The source directory of a project should encapsulate the source code in a clear python package definition. So that it can be used as a relative import as well as can be installed as a dependency for another project.

This enables the source code of the project to be used in the notebooks that would be used to further iterate on the existing version of the model, at the same time making sure it is usable in an ML pipeline as an importable dependency. This pipeline could be the pipeline of a different project or not.

Dockerfile

Dockerfile should be carefully crafted in such a way that caching makes sure that the change to source code triggers only the build for only one docker layer. All the efforts possible should be done to make sure that all the container image build results in a single layer being built.

The dockerfile should also be a multi-stage build that ensures that tests for the project are carried out and that a container image is only generated if the tests pass. The principle is related to Test Driven Development[1] because it ensures artifacts are only created after a test passes, ensuring functionality is verified every time a CI build is triggered.

Pipeline

Pipeline compilation should not need the installation of the source code or the project dependencies. This however becomes necessary when source code is imported in the pipeline code, and when such a pipeline is compiled this becomes a pre-requisite for the compilation (but hey, we are talking about utopian stuff).

Removing this requirement while keeping the benefits of function imports in pipelines (being able to write pipelines easily, an easy step on the learning curve for people new to the concept of pipelines) would be a life saver for CI times of compilation for pipelines and for base images to be used more widely.

The pipeline definition should be as simple as possible (possibly using a Python decorator over functions). This would enable people coming from non-infra/non-software-dev backgrounds to create complicated DAGs[2] with ease. On the other hand, this should not hinder the functionality of defining a pipeline component as a YAML file (like pro infra folks do.

Pipeline components should be reused as much as possible. Ideally, the forward the function of the training job is the new thing an MLE has to describe. A good catalog of reusable components would be available for MLEs to pick from without having to wonder about the internal details of the components (as these components would be maintained by MLOps folks).

---

[1] [1].
[2] [2].

Notebooks

Functions from notebooks should be importable in the source code src folder of the project, and further the src code of the project should be importable in the pipelines and components of the project.

This promotes bad habits of writing bad code in notebooks and then using it in the source code of the project, but at the same time, forces people to use functions in their notebooks. It's kind of a two-edged sword scenario, where people would write more functions and use pipelines easily (as adaptability could be a big issue when you introduce pipelines to a pipeline-less ML working culture), but at the same time, it would lead to non-production ready code in production. But the hope here is that the tests and other standards expected from an application will take care of this.

Ideally folks would use this until they are not sure if they are going forward with a pre-processing method (taking it as an example of the function defined in the notebooks) and as soon as the pre-processing method is finalized, they would either rewrite or re-vamp the function and move it to the src directory of the project and make it a part of the project source code.

K8s manifests

The above example assumes that you'd be using Kubernetes to deploy ML servers. How-ever, the same idea scales to any other kind of deployment. The idea here is that the manifest that control deployments would be saved in the project itself to make sure that there is only one place for a person working on the project to interact with. The existing CI/CD system directly picks up the git source of truth from the project repository directly.

There are a lot of tools that make it possible today, from GitHub actions to ArgoCD.

Infra-terraform

There are a lot of other infrastructure-related codes that are not related directly to the deployment of an ML project (in this case, we are taking Terraform as an example) but relate directly to the project itself. Most of the time, this is the part that needs to be bootstrapped when creating the project repository. This is maintained as a part of the git history of the project itself, and it can be changed/updated according to how the needs mature throughout the lifecycle of the project.

This idea might not sit well with the infrastructure teams having to maintain it, as they'd prefer to have this part under their maintenance to ensure relevant checks can be done before someone can provision any infrastructural component they wish to. This is, however, not the best way, IMHO. This needs to be tailored towards the experience of the user and not the experience of the infra-platform team. Also, there are ways to implement checks to keep control of what gets provisioned and how. There could be limits and checks implemented to make sure the CI jobs applying the terraform code need approval if they wish to apply a certain size of resources.

Workflows

The above example shows these workflows as GitHub workflows, but it could be any workflow, depending on the choice of CI. There are two most important workflows here; the first one would be the trigger to start a training job, and the other one would be to deploy a new revision of the model server.

Sometimes it makes sense to start a training job directly when a branch is pushed to the main branch for the dev environment, and the production training job could be triggered manually to make sure that we use our resources judiciously.

Depending on how the CD mechanism is setup, there might not be a need of the 2nd workflow, as the CD might be looking at the manifests directory explained above to make sure that any change to the default branch of the code would trigger changes to the resources directly.

Pre-commit checks

There's nothing new here, but making sure all projects have a sane standard set of pre-commit checks is the best and simplest way to elevate the quality of the code as a continuous task with every PR/commit that gets to the repository.

Dotfiles

Apart from the major chunk of the template described above there are a lot of .files that are needed a lot of times for various systems that teh code interacts with, a utopian scenario would be to just have a .mlops file that encapsulates all these files.

Also, this .mlops file should make sure to mention the stage of the project and the other options a project chose at the time of creation. Changing the variables in this file would make sure the CI creates an automated PR with the related changes that go hand in hand with respect to the change of the variable.

Let's take an example to understand this better: If I change the stage of my project to development from EDA. This would make sure that the Python source code structure gets generated for me to use. This makes sure that when a person is working on a project they are not overwhelmed with the files in the project, and it is easy for them to visualize the different stages of the ML project lifecycle and live through it.

Why is One Repository Per Project Used?

There has been a question that has come back to me time and again, which is that "If we have to create codebases so many times for ML projects, why not just have a mono-repository and maintain multiple projects at the directory level, we can easily get rid of having to setup x things that are needed to be done with every new project?"

This is a very valid question, and the reason I would suggest against it is to maintain the concept of namespace intact. A project staying in its own namespace makes so much sense with respect to organization, cleanliness, and other organization and maintenance aspects. If someone has pushed a credential to git by mistake for one of the projects, I would not be super happy making sure I am re-writing the git history for all the other projects. Maintaining a separate git history would also mean that one can easily make sure they archive a project or re-erect it from down under when needed. Some might even debate that having a mono repository would make it easy to integrate and deploy stuff, but I think I would respectfully disagree, as most of the CI/CD tools have very good support for multi-repository setup. One good example would be applicationsets[3] from Argocd.

When having a good template can give you the headstart you are looking for, there is no reason for one to consider a mono repository for ML projects that might have nothing to do with each other. It would keep the lifecycle of each ML project completely isolated from the other projects ongoing and being maintained.

CI

One of the most important and crucial aspects of an ML system would be the CI jobs that run when a code change is made to the ML project. This is also the most interacted part of the project.

Consider that you have a project that is in one of the deployment stages of the life cycle. In such a stage or in the development stage of the project, following are the kinds of changes a person would make to the codebase in their PRs:

When a new dependency is added to the source code, the CI should make sure that only one layer in the container image should be rebuilt, and everything else should get cached. The Dockerfile in the project should not take more than 5 min to build (considering that a CI run has already happened before to make sure that the cache could be used from either the previous builds or with the registry).

Making sure that the only new layer that is being built is the dependency installation, and all the other steps of the CI jobs are cached and tagged, can be a very difficult job to nail down.

When the source code of the project is changed without changing any dependencies, the CI job should only be building the layer where the source code gets copied in a container image or gets compiled to create a pipeline component. This can be difficult to nail, as to compile pipelines and components where source code from the src package is imported, you'd need to make sure that all the dependencies and sub-dependencies of the project are installed in the environment where the compilation takes place.

One common way to solve this issue is using a base image for a project, and to be honest, this works until it doesn't. So, when you are using a base image for your project,

---

[3] [3].

you update the last few layers depending on the changes made to the project. What goes wrong here is maintenance. Maintenance of the aspect of maintaining the CI logic, where someone has to write a code to understand which base image should be used for which branch of the codebase is in question here. Also, the best case scenario here would be that there is only some code change without changing the dependencies of the projects. In such a case, you need to make sure there is a good system to take care of, such as only adding the codebase to either the last layer of the container image or publishing the code to a Python code artifactory.

Then, at runtime (or any other time this code is needed, like pipeline and component compilation), you'd need to make sure that the code gets pulled from the artifactory along with the correct corresponding base image. Do you see how this is becoming a bit more complicated than it needs to be? Not just for the user to understand and work with the flow but also for the MLOps personnel to actually maintain a CI system that implements this (along with 9000 other corner cases that you'd encounter along the way; believe me, I've tried maintaining a similar base image system and I won't recommend it).

When a change is made to one of the pipelines of the project, in this case, the CI should make sure that all builds for the container image of the project are skipped and that only the compilation of the pipeline and components takes place. Ideally this should be one of the fastest case scenarios for the CI job to run.

Access Management

The next natural step would be to talk about CD, but before that there is another important aspect that we should be discussing, Access management. As ML projects have a shorter life cycle than other software projects, it is important that access management is seamless for any new repository that gets created.

Most of the ML projects would need the same base set of permissions, i.e., being able to pull specific data, read/write pipelines, artifacts, and components, and deploy services. When a project is created, all of these accesses should be automatically bootstrapped. It is very easy to set this up for an infrastructure team, as this is a common requirement for all other teams as well. ML needs some special accesses and would use this functionality more frequently than any other team, which would lead to MLOps and the Infrastructure team collaborating a lot with each other, and MLOps might even own and maintain a major chunk of project bootstrapping workflows as well.

A common thought when it comes to access management is what structure should be followed. Especially for ML projects that might be sunsetted soon, what structure would make the most sense.

One Service Account Per Project

A very common and easy solution could be to make sure that all the projects are just using one service account for all the access that happens on that project. This makes it very easy to remove/maintain and audit access to artifacts, as there would be a direct one-on-one relation between the project and the service account.

This strategy also makes the access provisioning workflows very easy to maintain. However, a drawback of this strategy is that it could lead to a single point of failure.

One Pair of Service Accounts Per Project

The drawback of having one service account is solved by having a pair of service accounts for each project. One with read permissions and one with write permissions, this makes sure that there is clear isolation of what a service account is authorized to do and this also aligns with the principle of least privilege.[4]

Whatever the strategy for access management, making sure that access provisioning happens automatically with a very well logged system, and a human audit is carried out on these logs every month is very crucial.

CD

Having an all time available continuous delivery system is the backbone of the core infrastructure of any organization. This is a core functionality and in most cases would either already be present or would be the responsibility of the central infrastructure team. If there are any tweaks needed for the ML deployments to happen magically, the MLOps people should own that piece of the delivery mechanism. This might be needed if the ML deployments happen using an inference serving framework.

Another often overlooked part of the delivery mechanism crucial for ML would be the deployments of production pipelines that predict on a schedule using an ML model. The delivery mechanism of such pipelines should be in a shape that the ML practitioners only have to worry about the definition existing in some git-controlled system, and then these pipelines just exist magically, work on a scheduled time, and deliver the expected results to the destination.

Another very important aspect of the CD system would be a very reliable alerting mechanism, if there is a pipelines that fails or a deployment that failed during the delivery mechanism or failed to be ready for serving requests, this crucial and time sensitive notification should reach the MLEs and on-call people (if the alert corresponds to a production workload).

---

[4] [4].

There are a few good tools that a lot of infrastructure teams rely on like ArgoCD,[5] CircleCI,[6] fluxCD[7] etc. Some of these tools might not be suitable for the pipelines kind of workloads but have proven to be the reliable good infrastructure as code continuous delivery mechanisms.

ML Process

One of the most important aspects is the ML process followed by practitioners. There's usually a very clear set of stages that a project goes through, but it's usually not understood in the same depth and with the same expectations by all the people working on the projects.

Almost everyone is aware that we'd do EDA, then train models, and then deploy one of the models to get predictions. The ML process is not this simple, though. As one dives deeper into the details of the process, it becomes more and more complex and confusing. There's not one way to do ML as well, there might be some divergences according to the organization, problem statement, historical context of projects, people involved and time available to deliver a working solution.

It is crucial to have a well-defined process and to have the same understanding of the process for at least all the ML practitioners. One example of this could be defining the ML project lifecycle and the expectations from all the stages. This would enable structure, knowledge sharing and common understanding of ML projects and their status across the organization by not just the MLEs but also the other stakeholders that are not familiar with Machine learning in extreme detail.

This touches a bit on the culture around ML in the organization, and the most important part is to **have a culture**. It's not about having the best culture and processes, and the ML process is something that needs to be continuously worked on as one goes forward with people/projects/needs and how the world evolves in the ML space. Just having a single and clear definition of the process of carrying out an ML project and the expectations from each step of the process would be a very structured approach to carrying out such projects at scale.

Dependency Management

One of the biggest problems with Python, in general, is dependency management, and this problem becomes a disaster when we talk about Machine Learning. ML is a very fast evolving field today. It is notoriously famous for not having ML dependencies that you

---

[5] [5].
[6] [6].
[7] [7].

can trust all the time. This can be due to the fact that a lot of ML originates from research based work (where the robustness of code is not of prime importance) or from the fact that something a few months old can be considered archaic already.

Python, being the language choice, doesn't really help. Python was made to be a language that one can easily prototype in. It simply isn't made to take things to production, and IMHO, we have developed a lot of solutions that help with managing the problem rather than solving the problem itself. There are a few initiatives where people have started to code in other languages like Rust, Go, and sometimes C++, to be honest, but those languages don't have the flexibility of Python and runtime.

There are a plethora of virtual environment management tools for python that one can choose from. These would make sure that you get an isolated namespace for installing a specific python version or dependencies, and then one can use pip to install dependencies of their project in an isolated namespace.

One issue in a ML python project is that you don't control the sub dependencies of the dependencies you are using, if you have pinned the exact version of your dependency, but your dependency has not pinned a version of their dependency, then you can't guarantee that you would be using the same code in every environment, as it would depend on the sub-dependency version that got resolved while installing your code.

This problem can be solved using better dependency management tools like pdm,[8] poetry[9] or pipenv.[10] Almost all of these tools work on the concept of locking all dependencies and respective sub-dependencies in a lock file, also making sure to note down the checksum of each pinned dependency version. This way these tools can ensure that the packages being installed are exactly the same that are intended to be installed.

The challenges begin to start when you are dealing with different operating systems (e.g., macOS versus Linux) or environments like Docker containers. These challenges primarily arise due to differences in how libraries, packages, and system dependencies are managed across platforms. Let's look at these challenges in depth is an in-depth look at how these issues manifest in different contexts:

MacOS versus Linux Dependency Handling

Differences in System-Level Dependencies
Python dependencies often rely on system libraries, especially when dealing with packages that have native (C/C++) extensions, like numpy, pandas, or matplotlib. macOS and Linux have different package managers and default libraries, which means the installation of system dependencies for Python packages can be problematic when switching between these platforms.

---

[8] [8].
[9] [9].
[10] [10].

Example: Consider installing a popular data science library like numpy or a machine learning library like Tensorflow.

On Linux, you might rely on apt (for Ubuntu/Debian) or yum (for RedHat/CentOS) to install system libraries on which Python packages depend.

sudo apt-get install python3-dev libblas-dev liblapack-dev

pipenv install numpy

On macOS, you might need to use Homebrew to install similar system libraries, but the library names and locations can differ:

brew install openblas

pipenv install numpy

Even though you're installing the same Python package (numpy), macOS and Linux require different system-level tools and libraries to handle the underlying compilation.

Issues with Binary Wheels

Many Python packages are distributed as wheels (pre-compiled binary packages) that are platform-specific. If a wheel is not available for your operating system, pipenv will fall back to compiling the package from the source, which can be troublesome if system dependencies are missing or incompatible.

For example, certain packages on macOS may not have pre-built wheels, requiring Xcode and additional system libraries that aren't needed on Linux.

Let's consider a real-world Issue:

Let's say you're working on a machine learning project that requires TensorFlow. TensorFlow uses specific hardware acceleration, such as CUDA (for NVIDIA GPUs), which is supported only on Linux. A pipenv environment created on macOS may install the CPU-only version of TensorFlow, but that same environment on Linux might attempt to install the GPU version. This can cause inconsistencies and incompatibilities when you try to reproduce the environment across platforms, even when using Python environment management tools.

```
pipenv install TensorFlow
# On macOS, you may get CPU-only TensorFlow
# On Linux, you may get a GPU-accelerated version, which needs NVIDIA drivers and
CUDA libraries.
```

Python and Docker Containers

Docker containers also provide an isolated environment for applications, but Python's dependency management with pipenv can run into a few issues when working with Docker.

Container Layering and Efficiency

Docker images are built layer-by-layer. This means that if you install Python dependencies using pipenv inside a Dockerfile, the entire virtual environment (along with potentially large dependencies) is baked into the Docker image. Each time the Pipfile or Pipfile.lock changes, Docker has to reinstall everything from scratch, making the build process slow and inefficient.

Example:

```
# Dockerfile
FROM python:3.8

WORKDIR /app

COPY Pipfile Pipfile.lock ./

RUN pipenv install --deploy --system
```

Here, every time you change your Pipfile or Pipfile.lock, the pipenv install step runs, reinstalling all dependencies, even if you've only made a minor change.

Platform-Specific Dependencies

Docker images often run on Linux-based containers. If you develop locally on macOS and your pipenv environment includes platform-specific dependencies (e.g., a package that compiles differently on macOS vs. Linux), this can cause problems when building or running the Docker container.

For example, a library like psycopg2 (PostgreSQL driver) may require different compilation flags on macOS and Linux. If you install it using pipenv locally on macOS, the resulting Pipfile.lock may not work correctly inside a Linux-based Docker container.

```
pipenv install psycopg2
# It works on macOS but breaks in Docker due to missing Linux libraries like 'libpq-dev.'
```

To resolve this, you might have to install platform-specific libraries inside the Docker container:

```
RUN apt-get update && apt-get install -y libpq-dev.
```

Inconsistent Virtual Environment Behavior

Pipenv creates a virtual environment in your local development environment, but in a Docker container, you generally want to install dependencies system-wide to avoid unnecessary isolation within an already isolated environment. This often leads to using the –system flag with pipenv in Dockerfiles, which can lead to inconsistencies when trying to mirror the local environment.

```
pipenv install–deploy–system
```

This command installs dependencies globally within the container, which is not exactly how pipenv behaves in a local virtual environment, potentially leading to subtle differences.

Multi-stage Builds and Best Practices

When using multi-stage Docker builds, Python dependency management becomes even more complex, especially when optimizing for image size and caching layers. In these cases, pipenv can introduce overhead by duplicating dependency installations or forcing you to manually fine-tune the installation steps.

General Docker Challenges with Python Packages

Binary and Platform-Specific Packages

Certain Python packages (like Pillow, an image processing library) rely on system-specific binaries. When you install such packages via pipenv on your local machine (say macOS), they will link against libraries available in macOS. However, when you build the Docker container (which typically runs a Linux distribution), these packages might need to be recompiled or fail to install without proper system libraries.

```
pipenv install Pillow
# Works fine locally on macOS but might fail in Docker due to missing system-level libraries
like 'libjpeg.'
```

In this case, you'd need to install the required system dependencies within the Dockerfile:

```
RUN apt-get update && apt-get install -y libjpeg-dev zlib1g-dev
```

macOS versus Linux: Python dependencies can behave differently because of differences in system libraries, paths, and package management tools. pipenv often needs extra configuration to handle these platform-specific dependencies. This leads to inconsistencies, particularly when switching between macOS development environments and Linux production environments.

Docker Containers: Pipenv has some limitations due to how Docker handles layering, platform differences, and system-level dependencies. Docker often requires using the `-system flag` with pipenv, which can introduce inconsistencies with local virtual environments.

In both cases, the challenge is in handling system dependencies and ensuring consistency across environments. The common solution to these problems is to use containerized environments more carefully, relying on explicit system dependency installation inside Dockerfiles and using cross-platform tools like pyenv[11] to manage Python versions.

Now that we have talked about so many problems with dependency management for Python let's look at what a dream scenario would be:

I would like to be able to just pin my direct dependencies, and not have to worry about what is happening underneath, I would like the sub dependencies to be pinned and locked automatically with my project.

Once the dependencies are pinned and locked, I would like to take that lock file to docker running on any platform possible and I should be able to reproduce the same dependencies on all these platforms.

If just these two simple sentences become a reality, the pain of Python and ML dependency resolution will cease to exist.

Notebooks/Report

Everyone loves coding in notebooks instead of writing Python scripts, and as soon as I got to use a notebook for the first time in my coding life, I was hooked on it. There was a point where I would always have a notebook on one of my browser tabs, ready to quickly try out something. This is the benefit of interactive kerl.[12] We can quickly and easily write code, iterate over it, and get results. It's a terminal and a text file combined into one. EVERYONE loves it.

EDA is the perfect stage of an ML project that fits this use case, and the world understands that. But the issue is that people use notebooks to deploy to production, as it's very easy to do so. This might even work at some scale, but a few things one loses with this are the ability to iterate over time and maintain that code over a long time. One loses the ability to pick up a piece of code after a long break and take a newer version of the code to production again. By the time you wish to do this, versions of dependencies have changed. Some code that was written dependent on some column name does not exist anymore. There is no log of how many changes were made to the code and what changes

---

[11] [11].
[12] [12].

were made. Essentially, no historical context of the code is available anymore. One can still have these issues with bad code written outside a notebook as well, as bad code is bad code everywhere, but notebooks promote such behavior.

The best case scenario for using notebooks would be to make sure that notebooks are only usable till the EDA stage of a project, and notebooks still are stored in a controlled environment, the code written is importable outside of the notebook (maybe store notebooks using jupytext[13]). At the same time, these notebooks are converted to a PDF type of report and are available in some report catalogs across the organization. This can be achieved using a lot of tools available today (one of them is jupybook[14]; another simple example could be using the markdown rendering feature provided by your choice of git cloud provider to publish these reports privately to folks in your GitHub organization).

This makes sure that their are easy ways for the notebooks to be used for 2 kinds of stakeholders for notebook artifacts, the one's that wish to look at results of the notebook and the one's that wish to use the notebook to make something out of it, or re-iterate on the result.

Artifact Storage

Managing artifacts and experiments effectively is important for maintaining a robust and organized workflow in machine learning projects, especially when scaling up to serve multiple teams or the entire organization. An artifact storage should be a centralized platform that provides a bird's-eye view of the machine learning projects across an organization.

There should be a dashboard that provides an overview of all projects, active experiments, their statuses (successful, failed), and key metrics, creating visibility across teams and departments. With such a platform, models, experiments, and data can be reused across projects. For example, a language model trained for one task could be repurposed in another task or improved by a different team.

Such a platform should also be consolidated with an Experiment management tool for tracking, storing, and analyzing all details of machine learning experiments, from the data used to hyperparameters to results. Every experiment in ML projects produces tons of data hyperparameters, metrics, logs, datasets, and model versions. Automatic logging of all this data would ensure that anyone can trace an experiment's outcome, understand the decisions made, and reproduce the results. This system should allow you to search for past experiments using filters like model type, date, or performance metrics. It acts as a searchable archive of past research, improving team collaboration. A few tools are available as experiment management tools that partially satisfy the above use cases.

Cold, Hot, and Warm Storage

---

[13] [13].
[14] [14].

Once artifacts are stored, the number of times they are accessed would be inversely proportional to the age of the artifact. Hence, just like logs are stored, it would make so much sense to make sure that the storage is classified into three different categories:

- Cold Storage: This is for artifacts you don't need to access frequently, such as old experiments, archived models, and datasets that aren't actively being used. It's cheaper but slower to access. Cold storage can be ideal for storing reports, old versions of models, or datasets that were used for completed experiments. This could be backed by long-term cloud storage solutions.
- Warm Storage: This is used for experiments or models that are used occasionally but need quicker access than cold storage can provide. For example, you may want to store models that are in the "staging" phase here. Warm storage could be implemented using solutions for medium-latency access.
- Hot Storage: This is for the artifacts and experiments you're frequently accessing, typically those related to production models or experiments currently in progress. Hot storage needs to be fast and highly available, often backed by SSDs or in-memory stores. Hot storage is typically used for production models, live datasets, and metrics logs that need real-time analysis.

In addition to storing models and metrics, this system could also be reused to store the notebook reports as an object that were discussed in the previous section.

In a large organization, access to artifacts, experiments, models, and even notebooks must be role-based and project-specific. One way to manage and provision this access could be managed in such a way that teams working on different projects have a clear overview of only the experiments relevant to their work. This could be enforced by linking access controls to specific repositories or folders in the versioned storage systems.

A well-structured system for artifact storage and experiment management can really enhance efficiency, collaboration, and reproducibility in machine learning projects. Tools like MLflow, Weights & Biases,[15] and AimStack[16] provide functionalities very similar to the one's described above. Such a system would act as a one stop solution for ML prationeers to look at all the projects that have been worked on, and the results of all the experiments. This would act like an ML Wiki[17] of the organisaiton.

---

[15] [15].
[16] [16].
[17] [17].

Data Consumption

To make sure ML teams can have a clear path to success again and again, the way data is consumed plays a crucial role. To ensure long-term success, we need to think beyond just handling data—we need to streamline how it's accessed, processed, and made available for ML operations.

First and foremost, we should be able to treat all types of data equally, regardless of whether it's structured, unstructured, time-series, or even more complex formats like vectors in the aspects of it consumed. This is key to making sure that one can enable MLEs to reach production as soon as possible with data not being a pain but a power for them.

Next, fetching specific data rows should be effortless. Whether you're dealing with a query from a business analyst or pulling training data for a model, the ability to pinpoint and extract exactly what's needed is vital. If fetching data is a cumbersome or slow process, it'll slow down everything else in the pipeline—so this is not just a convenience feature but a fundamental requirement for productivity.

Scalability is another important piece of the puzzle. We should be able to create and manage datasets at a terabyte (TB) scale and at a very small scale as well, ensuring that the tools and processes around this data can scale without bottlenecks. Whether you're training a model on a few gigabytes or a few terabytes of data, the infrastructure and processes should handle it seamlessly.

There has to be a clear point of interaction where data engineering ends and ML operations start. In an ideal world, this should be as frictionless as possible. Often, teams struggle because this handoff is blurry—either because responsibilities aren't clearly defined or the tools don't support a smooth transition. A well-structured approach would make sure that data is ready for use in ML, having passed through proper preprocessing, quality checks, and validation steps. Maybe having a feature store as a clear layer that handles all kinds of data in between the data engineering and ML teams can be a good system that would solve this problem with elegance.

A well-maintained data catalog is the backbone of a strong data ecosystem. There should be a centralized and easily accessible inventory of all available data sources, including metadata that clearly defines what each dataset contains, how it's structured, and where it comes from.

Beyond just serving as a reference, the catalog should also be proactive, alerting teams to any data drifts—a common issue when working with real-world data. If the distribution of data changes, it is very important that the data consumers get this information so as to make decisions on how an ML model performance would be affected by this. Having these alerts ensures that the teams can act in real-time to address issues before they snowball.

Even though one might have the alerts set on the data source and catalogue, it's very important to make sure there's a quick check that happens before we start to use that data (say a pre-training-check for data).

That brings me to another key point: drift alerts should be tightly integrated into your workflow. It's not enough to know that your data has shifted; you need to have a clear and direct way to receive these alerts and hook into your monitoring systems. Whether through email, dashboards, or automated notifications, staying on top of these changes is essential for maintaining model accuracy.

Finally, there should be a one-stop solution for exploring, accessing, and monitoring all datasets. Think of it as the data hub for the entire organization. Teams should be able to quickly explore what's available, understand how to work with it, and be notified when something important changes—be it schema updates, new data availability, or potential quality concerns. This hub acts as a single source of truth, fostering both collaboration and operational efficiency across departments.

Training Infrastructure

This can be a very intense infrastructural journey and work to put in, as we are talking about the most important aspect and stage of an ML project, Training a model.

It's not just about having powerful hardware, but it's also about creating a system that is flexible, efficient, and reliable for every stage of the model training process. If we dive deeper into the training stage a bit more, we can look at the key aspects of a scalable model training infrastructure, breaking them down so that we cover all the technical details while staying clear and approachable.

Fast and Flexible GPU Provisioning

When it comes to training machine learning models, GPUs (and sometimes TPUs) are the most precious resource (especially with the onset of bigger models coming into existence). It's very important to have them available in this GPU-hungry world, but it's not enough just to have them available; they need to be provisioned in seconds. If you've ever had to wait around for a GPU to become available or watched as resources sat idle, while your team waited, you'll know how frustrating that can be. This becomes even more frustrating if you have the GPU resource, but the container image is taking time to be pulled in, or you are waiting for the data or the model to be loaded from another part of the globe (regions for all cloud resources matter). You want a system where the moment you start a job, the necessary GPUs are ready for you.

But here's where it gets even more practical: GPU sharing. Not every job needs an entire GPU. In some cases, you're only using a fraction of the GPU's capability. So, it makes sense to have infrastructure that supports fractional usage, letting smaller jobs share a GPU without wasting resources. This helps you keep costs down and ensures efficient use of all available resources. Most of the Kubernetes providers make sure they provide some sort of GPU sharing available within the cluster, either time-based sharing or priority based sharing. This optimization should be already available in the training infrastructure.

Reliable and Reusable Training Pipelines

Training pipelines are the backbones of model training workflows. They might be handling every step from data preparation to model training, evaluation, and eventually deployment. You need a pipeline engine that is robust and reliable. You don't want a system that breaks down in the middle of a job, especially with large-scale models that can take hours or days to train.

A good pipeline system should also support caching. Think about it: many steps in a pipeline don't need to be recomputed each time you run the model. If you're reprocessing the same data again and again, that's a huge waste of time. Caching allows you to skip steps that haven't changed, significantly speeding up your workflows (It is really important for MLEs to be able to specify if they do not want to use for their cache step as well).

Another critical feature is pipeline reusability. MLOps teams should maintain a set of well-tested, reusable pipelines that any ML practitioner in the organization can pick up and use. These pipelines serve as building blocks, so you don't have to reinvent the wheel for every new model you're training. You just take what's already there and adapt it as needed. These should be made available as Lego blocks, and the ML practitioners should just be able to fit them in their building as blocks.

On top of the cache, the ability to schedule pipelines is a must. Whether it's for daily model retraining, weekly batch jobs, or real-time streaming tasks, pipelines need to be scheduled and executed automatically, without any manual triggering. This can be pivotel in making sure that one can automatically update a model in production provided it passes the evaluation after training and is proven to be better tha the existing model.

Monitoring and Alerts

No one likes to babysit a pipeline run, so having a solid alerting system is essential. Whether a pipeline fails or completes successfully (maybe not on this one), you should get a notification immediately. This allows you to jump in and fix problems as soon as they occur, or move on confidently, knowing that everything is working as expected. Infrastructure folks usually love alerts, so they would be super happy to know that you are monitoring pipelines.

Apart from the alerts that need to be delivered to the pipeline authors (usually ML practitioners), there needs to be overall monitoring of the pipelining engine to understand how many pipelines are being run at a time, how many pipelines fail every day, how many pipelines fail due to a system error on the pipelines. These errors and monitoring dashboards are for the MLOps team to observe and aim to take these metrics to the positive side. These clear views of failure rates make it easier to pinpoint recurring issues in the system. Ideally, less than 1% of pipelines should fail due to infrastructure problems, and if they do, there should be clear logs and metrics to help troubleshoot.

Versioning and Reproducibility

One of the key aspects of a reliable infrastructure is ensuring that everything is versioned properly. Your pipelines should be defined in Git, right next to your code in the same Git repository, ensuring that any changes are tracked. This makes it easy to reproduce past runs or roll back to previous versions if something goes wrong. It also makes it easy to create, update, and remove from the single point of interaction in the code of the repository.

Moreover, if a pipeline fails due to system issues, it should be able to resume from where it left off. If a model training job that has been running for 12hrs only to fail at the last step, instead of restarting from the beginning, the infrastructure should allow you to pick up right where it failed, saving time and resources.

Another sweet functionality is to be able to trigger pipelines from the git interface itself, rather than having to go to another system and create runs. Simple automation can be built by MLOps to trigger runs on pipelines by adding PR/issue comments (with necessary parameters for the pipeline). This can make sure that ML practitioners can easily create a pipeline, run it from git, get the results back as a result in the PR itself, and make it easy for the reviewer to understand what the pipeline is supposed to do. All of this can be done without having to leave the git interface at all.

Pipeline Visualization and DAG Management

Pipelines, especially complex ones, often have multiple steps with various dependencies. To manage these, you need a system that allows you to visualize the DAG (directed acyclic graph) of the pipeline. This visualization is crucial for understanding the flow of tasks, spotting bottlenecks, and debugging issues.

As pipelines grow more and become complex, it's helpful if the system can simplify the DAG for you. Sometimes, the complexity of a pipeline can be overwhelming, so having a tool that breaks it down into smaller, more understandable components is a huge advantage. Sometimes the way a DAG is defined could be not the most optimum graph, and hence such simplification before running the pipeline can make it easy to execute on the infrastructure.

Scalability and Stability

The infrastructure should be able to handle a variety of different workloads, from quick pipelines that run in seconds to long-running jobs that could take days or even months to complete (someone is training the next GenAI feature of their app). The system needs to be stable enough to ensure that no matter the pipeline size, it doesn't crash or lose progress. Scalability is critical, so whether you're running hundreds of small jobs or a few massive ones, the infrastructure should manage them smoothly.

Another important aspect of scalability of the application would be the shared storage, it should be easy to set up and manage. Multiple parts of the pipeline should be able to access the same storage, removing the need to duplicate data. This is especially important for teams working with large datasets or complex models.

Seamless Deployment Integration

Training pipelines shouldn't stop at just training a model. Ideally, the infrastructure should be built so that after training is complete, the pipeline automatically triggers model deployment. However, before deployment happens, there should be an evaluation step that compares the new model against the existing one in production. If the new model passes certain performance benchmarks, only then should it move forward to deployment. This ensures that only models that are demonstrably better with respect to the evaluation metrics defined are the ones that replace existing ones.

There should also exist re-training pipelines that should be available with the simplicity of only defining the re-training schedule, and metric to evaluate, everything else should just happen automatically as a magic wand switch and flick.

Data Quality and Monitoring

Every training pipeline should include data validation checks to make sure the data is following the expectations, is complete, and ready to use, as data is the foundation of any model, and ensuring its quality is paramount. (Remember, garbage in, garbage out[18]). This can help prevent issues like missing data, incorrect formats, unexpected distributions, or outdated inputs that could lead to poor model performance.

In essence, building infrastructure for model training is all about making things fast, reliable, scalable, and repeatable.

---

[18] [18].

Vector Storage and Retrieval

This could be an optional topic for you, depending on what kind of ML problems are usually solved in your organization. A lot of times, this might not even be something relevant for your ML practitioners, and if that is the case, you could look over this for a while. There are a lot of tools available today to store and retrieve vectors, and almost all of the tools have their pros and cons that they would be good for. Some scale better at large sizes of indices, some perform better at small scales with a very efficient retrieval, and some perform better when retrieving similarity over certain distances.

Flexibility with Vector Sizes

A storage system should be able to handle vectors of different sizes and types. For example, vectors from an LLM might be 1024-dimensional, while image embeddings could be much larger or smaller. A vector storage system needs to support such a wide range of vector dimensions without any performance issues or compatibility problems. This flexibility would ensure that it can handle a variety of tasks, whether you're dealing with small embeddings or massive vectors.

One of the primary use cases for vector storage is performing similarity searches and retrieving the most similar vectors based on some distance metric (e.g., cosine similarity, Euclidean distance, etc.). The system should support fast and efficient nearest neighbor searches, whether you're working with 100 vectors or 100 million vectors. The ability to adjust or swap out similarity metrics based on your specific use case is also important. So, the system should be able to optimize the storage according to specific distances and metrics that correspond to the required retrieval method used and the use case.

When storing vectors, latency is a key factor, especially when your ML models are serving real-time predictions or recommendations. If you're building systems that are recommendation engines, and you need to serve similar results to an existing product, having low-latency retrieval is crucial. The faster the vector database can perform lookups, the better your user experience will be. For real-time applications, your system needs to be optimized to deliver quick responses, even under heavy load. For some use cases, it might not be possible to predict on the fly, but if the possible number of inputs is limited, one could batch predict and store the results in vector storage, making sure that the predictions would be available at very low latency when retrieval is needed.

The vector storage should be able to horizontally scale as well, as the size of the vectors keeps growing. It's especially important in organizations that deal with millions of vectors or rapidly growing data pipelines. A scalable, distributed architecture will help you handle this growth efficiently.

Vector storage systems should integrate smoothly with existing ML pipelines. Whether you're training new models, updating embeddings, or running inference, the storage system must fit neatly into your workflows. It shouldn't require complicated workarounds to ingest or retrieve data, and it should support automation as part of your model training and deployment cycles. Seamless integration helps avoid bottlenecks and ensures that ML practitioners can focus on their models instead of dealing with infrastructure headaches.

There exist some vector storage systems that have a very efficient retrieval mechanism, but the only drawback is that if you need to add more vectors and wish to search in the index, you'd have to completely reindex the vectors. A lot of times, reindexing the complete vector space is not feasible. In an ideal vector storage system, you should be able to incrementally add new vectors without having to rebuild the entire index from scratch. This saves both time and compute resources, especially in dynamic systems where new data is constantly being ingested, and retraining or updating happens regularly.

As the number of vectors being indexed increases, the system should auto-scale based on this load. This ensures that the system can automatically allocate more resources (compute or storage) as the needs of the workload grow, without any human intervention. Auto-scaling allows your infrastructure to grow dynamically and meet changing requirements, ensuring that performance remains stable no matter how many vectors are being added.

As vector storage systems evolve, you'll need to migrate to newer versions or even different tools. An ideal system should make version migrations as easy as possible, ensuring that you can upgrade without causing downtime or needing to redo a large portion of your work. Also, keeping track of different versions of vector indices and models is essential for reproducibility and for rolling back in case of failures.

The infrastructure for storing vectors should be future-proof in the sense that if you need to migrate to a new system or update an existing one, it should be a smooth process. The system should also allow you to quickly adopt new technologies or upgrade without having to rebuild the entire architecture from scratch.

Serving Infrastructure

The second most important infrastructural component would be the serving infrastructure beneath the deployments or pipelines. This component has to be the most robust piece of the MLOps puzzle. If the training infrastructure has issues, one cannot produce a new model, but once the serving infrastructure has issues, you are not serving live requests. It doesn't get worse than that. There have been instances where services being down for a few hours have almost bankrupted companies as well.

We would majorly be focusing on the serving infrastructure of model servers rather than serving pipelines. Most of the infrastructural work put into training pipelines can be directly re-used in the prediction pipelines as well. Just that the output of a prediction pipeline might be some batch prediction file being saved in a location. This would be the best use of the existing work that has been put in to make the training infrastructure reliable.

Diving a bit deeper into the model deployments, a basic necessity is that once one creates a model server, one should not have to worry about configuring alerts for the server at all. This should be a core functionality of the serving infrastructure that would take care of the service discovery and make sure that alerts are scrapped form a pre-defined endpoint, also to generate alerts on health status of the application automatically.

At the same time, there should also be alerts defined on the quality of prediction of the model, which would be reused every time there is a new model put into deployment. In essence, supporting smooth integration for A/B tests. Along with this, the serving infrastructure should provide gradual rollouts for new model deployments.

Another very important aspect of the serving infrastructure has to be simplicity and ease of use, and the process should be very smooth. Ideally, one should only need to provide the model path from the storage, and everything should just magically happen.

Pulling the model from the storage, understanding the type of model, using the respective image with prediction logic, translating this to a machine on infrastructure, testing the model deployment with existing evaluation metrics, once the model is ready, running a small post deployment test and mark the model ready according to the default rollout strategy should just happen.

Just point the system to where the model lives, and it should handle the rest.

Obviously, there also needs to be some flexibility for people who have more complex needs, and those should be options that the model author can configure. There has to be an easy way to add custom forward functions, pre-process and post-processing functions, and define scaling and rollout strategies. Another ideal scenario would be where one can specify the required RPS, and the model scaling is automatically taken care of. In essence, while it's nice to have the "plug-and-play" option with minimum configuration needed, there should also be room for custom setups when needed.

Another really important thing is making sure the system can handle migration. You don't want to be stuck on one platform or one type of infrastructure. There will come a time when you'll have to migrate, and it's very important to have that at the back of your mind and make careful choices that would support this. If you need to move your models from, say, your own servers to the cloud or between different cloud providers, the model-serving system should make that as smooth as possible. You don't want to be in a situation where migrating means redoing a ton of work, or worse, causing service downtime.

Once the model is deployed, you can't just assume everything will work perfectly. That's where post-deployment tests come in. These tests are run after the model is deployed but before it starts handling live traffic. You'll want to feed the model both sane inputs (the kind of inputs you expect from real users) and insane inputs (crazy or unexpected data) to make sure the model responds in a way that makes sense. This helps catch any weird bugs or edge cases before the model starts serving real predictions. Only once the model passes these tests is it considered ready to serve live traffic.

It's very important to store all the inputs that the model sees and the predictions it makes. It's not just about keeping a log. It's about understanding how the model behaves over time and using that data for retraining. If the model starts making strange predictions or if the data changes in some way, you want to have a record of what's been happening. Also, storing this information helps you continuously improve the model by feeding it back into the training loop.

In summary, a good model-serving system should be easy to use, flexible enough for custom deployments, able to scale and migrate across infrastructures, and thorough in its testing and logging. This way, you can confidently deploy models that perform well in real-world situations without worrying about whether the system can keep up.

Incident Handling

Even if you do everything right, there will be an incident, and it will get your production servers down. Let's accept this truth, and let's prepare for this day in advance (while hoping that this day never transforms into reality).

There has to be a clear definition of who is going to be on-call. There could be 2 kinds of production incidents that happen with ML services, one of them could be a generic software engineering issue, and the other one could be if a model suddenly starts giving predictions that don't make sense.

There have to be clear definitions of what needs to be done when an issue occurs with an ML service. This closely aligns with a question if ML practitioners should be on-call? This could be very difficult to answer, as this question refers to people.

One good reason in favor of having ML folks on call is that in case something goes wrong there is someone that understands the ML lifecycle from end to end. This could be very useful in making sure that the issue gets fixed as soon as possible.

However, on the other side, in case the issue is with the software engineering aspects of a production workload, any on-call engineer can help in solving it, and if it's an issue of ML model predictions, apart from rolling to a previous version of the working model and checking the double the distribution of inputs for the model is getting for prediction. The ML practitioner would also take a few days to analyze the model and fix the issue for the model.

These two steps could be made using a clear definition of the runbook when a model gets an alert.

I would still vote toward having at least one ML/MLOps person on call all the time. To make sure that if something goes wrong, we would be able to resolve the issue as soon as possible.

Another important aspect is to make sure there's a postmortem that happens once the incident has been taken care of. These post-mortems would focus on jotting down the points that would have led us to avoid this disaster in production once this list is jotted down with the help of system owners and the first responders of the incident. The system owner team is to make sure that these points are taken care of so that a similar incident can be avoided in the future.

Red Team

Red teaming[19] is a concept that comes from the security industry, which is usually mentioned as the a team that would make sure to try and find security flaws from the perspective of a malicious user, and provide a report on what flaws should be fixed in the product.This team would take one of the most important decisions for a product, the go-no go decision, essentially understanding if the product is ready to be rolled out to users or not.

Such a red team is also required for ML teams before a model is set to be released in the wild. The world is not new to ML model fiascos, not just from small-scale organizations but also from a lot of the biggest names in the tech industry. Hence, before a model is decided to be put into production, it is important to let the model go through a red team check. The goal of this team would be to make sure they ask all the possible questions that can break the model or lead it to give predictions that won't make sense. This team should be very pessimistic in its approach while reviewing a model. Neither should this review be taken against the authors/collaborators of this model, nor should all the issues pointed out be fixed before taking it to production. The idea is to make sure that one is aware of what can go wrong and how it can go wrong with the model. This report would compromise issues that can either be fixed or the model to be protected against such attacks.

Also, the unfixed issues should also be included in the on-call run book of this model deployment. So that incase of an issue remotely close to the one pointed out by the red team, the on-call team would be aware of what to do to mitigate the issue.

Usually, a red team should definitely include 1 MLOps Engineer, 1 Site Reliability Engineer, 1 Software Engineer, and 2 ML Engineers. This would ensure that the red team's suggestions consider multiple perspectives on the same deployment. This red team should also come to a common consensus on whether this model is ready for being deployed in production (no matter what product needs). If there is an immediate need for the model to be put into production and serve live traffic, the major issues pointed out should be solved as soon as possible, and the other issues should be detailed in the runbook, along with suggested fixes and the symptoms.

Along with answering all these questions there should be a score the red team has to assign to the ML system with respect to the level of effort needed for maintaining this

---

[19] [19].

system, this would make sure the future planning for work on the system can be planned out carefully with respect to the return of investment in the system.

Components of an ML Platform

The diagram below mentions the core components of any MLOps platform and the teams that should be responsible for maintaining parts of the platform. All the components in the diagram need not be described explicitly as they have been discussed as a part of this book in different chapters and sections (Fig. 3.1).

Question to Assess Your MLOps Maturity

Now that we've gone through a good list of utopian standards for ML systems, it's crucial to understand how one can measure where they currently stand and how far they are from achieving that ideal scenario. The following list of questions would not be an exhaustive list but is a list of questions that ranges through various levels of understanding and maturity. It's important to note that there is no right answer to these questions. These questions are made in a way that they have to be assessed periodically, ideally annually.

The way an ML/MLOps team is moving on an assumed scale, based on their answers, will clearly reflect how progress is being made. These questions are meant to be eye-openers for individuals and teams, encouraging reflection and triggering insightful conversations. They are designed to prompt thought, inspire action, and guide continuous improvement.

Let's look at these questions with some considerations around the topics that they relate to:

Data

Do you collect data for the future?
In simple terms this would mean if you collect data even if there is no path for clear usage or need. There has to be some notion how this data would or could be useful in the long term, but if a clearly defined path for how this data would be used considering the current tech is not in sight, do you still collect and store data in an organized fashion?

Do you prepare datasets before you see a clear path of usability?
This question is a bit similar to the first one but takes us a bit closer to ML by asking if you create datasets using data that you have collected or continuously collected. This would seem useless at first, but hear me out (or read me out). A lot of times, you wish to

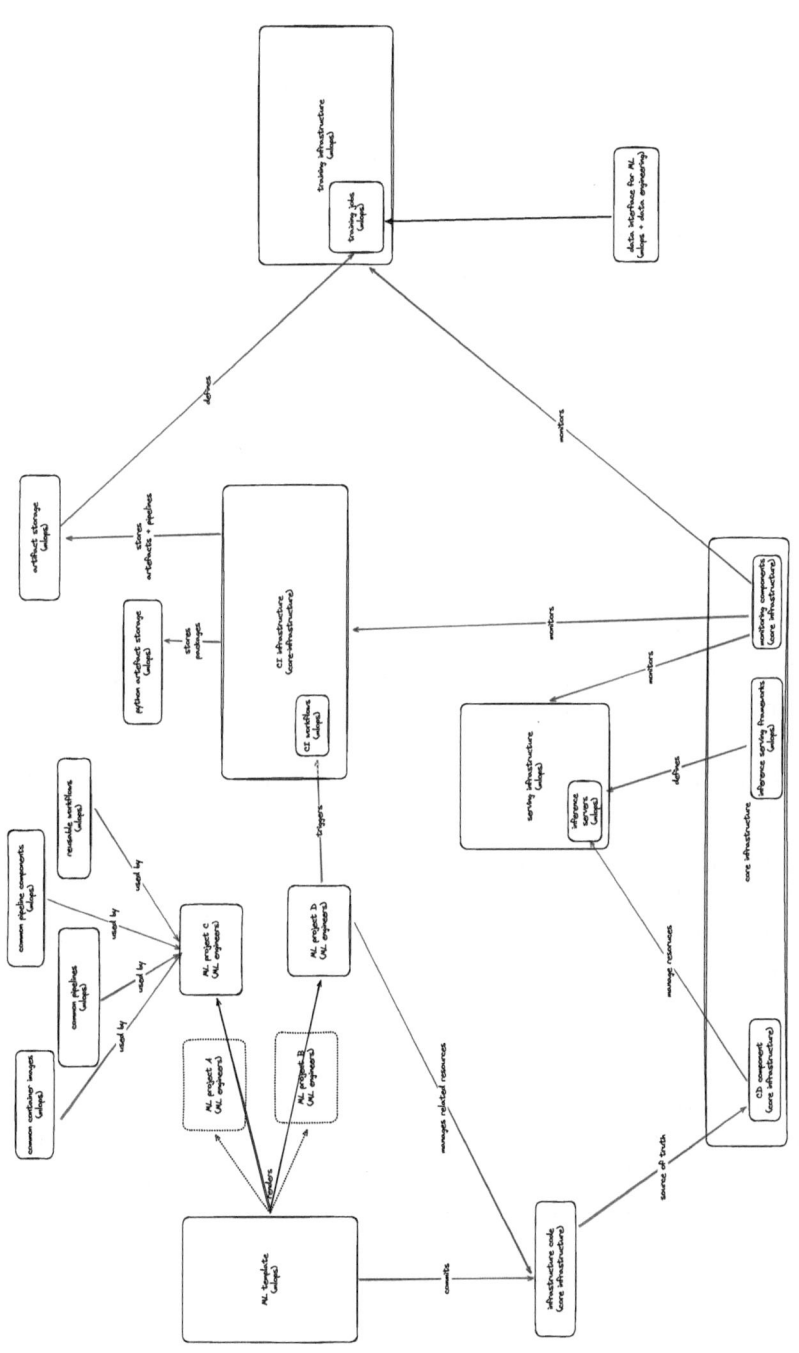

Fig. 3.1 Components of an ML platform

solve an ML problem, and you see a clear way to create a model that would work, but as you did not have a data collection strategy, you'd have to start by re-building the wheel. A good example could be if you are working at a firm and you wish to create a model for predicting churn, but due to the lack of data collection strategy, there are a lot of data points missing (like frequency of logins, support ticket ratings for users, etc..). Only if a dataset called users had existed and had been prepared iteratively over a period of time could building such a model have been an easy job.

Another good use case for these datasets, could be to open source them and give back to the ML community. This helps not only in contributing to the common good of the ML world but also helps a lot in marketing our organization as one of the known names in tech.

Is it fairly easy for you to fetch a part of the data and start with EDA?
Would you say it's super easy for you to just hop onto a notebook, call a data source, and use a subset of data from the source to carry out EDA? And if it's easy for you to do so, can you do this with all the data sources in the organization? This question is also related to a data catalog, where any data/ML practitioner should be able to discover and pick a subset of data, explore, and find potential gold in the data mines of the organization.

What is your default data collection strategy?
This question is meant to make you think about what the default data collection strategy at your organization is, and if it doesn't exist, should you? In specific cases, there's obviously consideration to be made about what to store and how to store it. Unless a special circumstance has happened, what is the default collection strategy known to everyone who works with data and products?

Do you store as much data as possible? Do you go for storing no data? Do you go for storing data only if PII doesn't exist?

All these questions are something you and the complete organization should be clear about and the strategy should be clearly documented and communicated across.

How easy is it to create multiple versions of the same dataset?
When you are working on EDA or creating preprocessing pipelines, is it easy for you to create multiple versions of the same dataset? This question dives into the aspect of versioning a bit, but, in essence, it tries to talk about whether there are clear tools and processes for creating multiple versions of the same dataset. Also, this touches on how to make sure that the infrastructure lets you create datasets and also supports such operations.

For the sake of an example let's say that an organization creates datasets as a directory that holds structured and unstructured data, and then the directory is serialized in a dataset type called directory.dataset. This dataset file is treated as a version in itself, this was done to make sure that once the datasets is pulled to the machine it can be used to saturate all the machine resources in one go.

Now, if I wish to change a specific column of data or if you want to change all the images of the dataset to sepia mode and black and white mode, Due to the packaging and the versioning method used here, you'd not call it a very easy process to be able to create a new version of this dataset, without recomputing the complete dataset from the data source. Had the versioning strategy of the dataset been on the objects level instead, there was a possibility of only re-computing the specific columns and objects that you wish to change, and other objects/columns of the dataset could be reused.

Another important aspect is that there would be different kinds of datasets depending on the kinds of data sources and data types that are included. This question is asked in general about all the datasets that your ML practitioners work with or would work with in the future.

How easy is it for you to jump between different versions of data?
Now that we've talked about the ease of creating dataset versions, it is very important to understand the need for usability of the same. Once you have multiple versions of datasets, how easy is it for you to change the version of dataset in your EDA jobs, notebooks or training jobs?

Are you confident that your data usage is compliant with regulations?
How confident are you about holding a data audit in your organization? Are you super sure that every corner in your organization of the data being used is compliant with 100% coverage of compliance across eerie data consumers and producers?

Can you rollback to the 7th last dataset version you used, and take your models to production with it?
This is one of my favorite questions. Let's take a scenario where you are doing EDA, and trying out different versions of a dataset. While you are comparing a lot of different versions, you realize that the 7th last version of the dataset you tried is the one that can give you the best result.

Would it be easy for you to trace back to that version of data, and code, and continue from that point onwards to create a training job, and deploy a model to production. This is a bit dependent on how organized you are when working in notebooks and lets you ponder a bit upon the experiment tracking and traceability of datasets and the code that uses it.

Do you use special DBs for vectors/images/videos?
This is a very straight forward question, if you deal with different kinds of data types, do you have specialized storage solutions for these data types or are these stored in a usual object storage solution.

As no question has a correct answer here, it is very important to understand and evaluate the current storage and retrieval method of different data types and what the point is when you should switch to something specialized. The switch might be related to the use case you wish to serve, or could be related to the scale of the data stored, or could be related to how fast you wish to retrieve different data types.

Pipelines

How good are your failsafe mechanisms for automated workflows?
It is very important to have an estimate of how reliable your automated workflows are, and top of the reliability of the workflows, there need to be failsafe mechanisms for workflows. These failsafe mechanisms could be as simple as automated retries and error handling mechanisms. Or could be something more complex on the lines of checkpointing and caching for individual components of the workflow.

Apart from these failsafe mechanisms, monitoring and alerting is another very important failsafe measure to have in place as a part of the fallback procedures for workflow failures.

How difficult is it to add a new step to an existing production workflow?
This could be a very good measure to understand the maturity of your pipeline orchestration. The time taken to make it to production can be a very confusing metric to look at, as the faster you can make a change to production, the better, but this metric cannot be looked at independently. This metric should always be looked at by some other metric representative of the robustness of the production workloads (maybe the number of issues in production per month). The optimization of both these metrics should happen by treating them as complementary metrics, just like precision and recall.

How do you maintain the versions of workflows?
The prime motive of this question is to make you sit and look at the versioning system of your workflows, if it exists. If this is a clickOps situation for you, there's a lot of room to grow towards a more mature versioning solution.

How do you maintain dev/prod parity in your workflows?
Slightly connected to the previous question, this one touches on the points of environment isolation and the development of pipelines.

Do you use a single machine to train models?
How many machines does your training job use? Is it a single machine that is kept in your office? Are you using the cloud? Is there some sort of orchestration that leads you to use multiple machines to finish your training job?

What is the time you will take to replace an existing deployment with a new version?
The time taken by an ML practitioner to replace an existing deployment with a new version is a good indicator to understand how a lot of things are coming into play. This represents if it is easy for the practitioner to perform operations throughout the ML lifecycle. The more complex the model deployment in question, the better the measure and maturity of the ML system.

If your training machine fails, can you easily resume the workflow at any checkpoint?
What is the checkpoint mechanism in place for the training pipelines? Is it possible to make sure if a training job that was supposed to run for 30 days and it fails in 25 days, one should be able to start the training from the last epoch and checkpoint, rather than losing on all the previous days of progress.

How granular are your workflow checkpoints?
If you are choosing to have checkpoints, how granularly are you storing them and how granular should they be? at the top level these could just be the caching at the components of a pipelines, the second layer would be caching or checkpointing at the level of each epoch of a training job. Practically this would be the only level till which checkpointing and caching should be implemented, but can be made more efficient at all times.

Are you using distributed training workflows? can you federate your training workflows?
Are your training jobs capable of handling ensemble models? Are you able to scale horizontally and create distributed training jobs? Distributing your jobs can drastically reduce the time taken for training models. If you are not using distributed jobs today, it might be a good question to ask why not, when you can start, and how it would help you. Using federated learning could be another big possibility to explore.

Do you have a way to seamlessly track experiments and generate meaningful assets/reports?
Having talked about the distribution of training jobs, and versioning of pipelines. There needs to be a very well ordered and managed experiment management and artifact storage systems that can manage the chaos of versions, containers, experiments and artifacts generated by each pipeline.

Do you have the ability to trigger other workflows if the parent workflow fails?
Another failsafe mechanism is to be able to run other pipelines on failure triggers of certain key workflows. This would not only enable to run fallback workflows but would also trigger other workflows that start re-trainings or start a lockdown schedule. Even if you have the ability to trigger such dependent workflows, the ease of being able to do so is another ladder to climb.

Versioning

Do you *version your data?*
If you do, is your data versioned at the object level or another level? What are your thoughts about versioning unstructured data?

How do you work with versioning w.r.t GDPR compliance?
Data versioning, or any other versioning for that matter, is about creating and storing multiple copies of an asset. However, GDPR/CCPA compliance forces organizations to delete all PII data if a user requests it. One clear way to store work around this is to remove Named Entities from your data. But a big thought bubble here is how you work with compliance with the versioning and storing of data. Is the best way to redact all the data before even storing it or before creating datasets out of the data, or is it better to redact data as per the requests received in all the versions of the data created?

How do you tie the versions of all kinds of resources together?
If you are versioning different artifacts, how are you tying the multiple versions of ML artifacts, pipelines, and code together? Also, how are different versions mapped to each other, and what kind of mapping exists between these artifacts? (is it one on one relation, one to multiple).

How do you maintain versions of your code/experiments/data/deployments/workflows/ assets?
This forces people to have a clear versioning strategy for all the ML assets. Having this clear definition and following it to register all assets is a sign of a mature ML system.

If you get a confusion matrix, would you be able to trace the data version that was used to generate it and the version of model deployment this plot will correspond with?
This is another one of my favorite questions, as it leads one to think about a very simple example that they can come across while browsing through an ML asset management tool. If someone is able to confidently answer a YES to this question, I think they should talk to the MLOps people who work at their organization and treat them to a big, hefty lunch.

Tests

How many tests does your first version of a production deployment go through?
This is a very important metric to measure as well as a very interesting thought, to understand how sure are you with what you ship to production. I've had experiences of seeing models in production that have never seen a test in their life. Those situations are as good

as driving my car blind, or taking a flight when no maintenance check has been done on the plane.

Do you have a clearly defined process on how to maintain models in production?
When something is in production, there should be clear definitions of how a production should be iterated, when it should be iterated, and how it would stay in production for a long number of years to come.

How many checks would a code change have to pass to make it to production?
This might seem like the same question as the first one, but it's not. This is about how many tests a code change done to the source code of the ML project would have to cross to end up as the live server. On top of how many, a good thought exercise would also be determining what kinds of tests this code change would go through.

If you get an automated PR from dependabot,[20] *how confident are you to merge the PR if the tests/CI checks are passing?*

This is another question of my taste, and the ideal scenario would be to automate such PRs to be automatically merged if the tests have passed. One should be confident that a minor version of a dependency change with passing tests would not bring down a production system.

What's your strategy to avoid security vulnerabilities in your code while sticking to specific versions?
One of the biggest challenges with ML dependencies is that X works with Y version, but does not work with Y + 1, but Y has a major security CVE tied to it. How do you handle such situations? A few options include making sure to either contribute upstream to packages, use a forked version of the dependency, or convince the security team of your organization that everything is fine. Whatever option you choose, it's super important to have a clear process around this and navigate such situations with ease.

Do you have tests on your data sources and model predictions?
When someone says the word test, we usually think about code, but having tests on your data sources and your model predictions is equally important to have a stable and maintained ML ecosystem.

Deployments

How much time does it take to spin up a new kind of deployment?

---

[20] [20].

This is a very important measure to understand how easy it is for someone to spin up a completely different kind of deployment. This can be a very good measure to understand how opinionated the ML system is and how many outliers it can handle.

How many different kinds of ml deployment do you support in your organization?
With the ever evolving ML scene, the different kinds of ML deployments supported by your ML system are a very good measure of how mature and evolving your setup is. One of the most asked for deployment/model types would be RAGs, as an example. If your MLOps team supports the ability to easily put up batch, live, and RAG prediction servers, this number would be three. At the same time, it is important to look at what other deployment types could be included in this list.

Do your ML deployments auto-scale?
This might sound a bit basic, but auto-scaling deployments would definitely be a default confirmation from the infrastructure in certain ranges. Make sure that the ML deployments auto-scale according to the number of requests and time taken by the prediction time for a single request. These scaling configs should also have a cap as ML deployment could, a lot of times, be very expensive in terms of resource consumption directly relating to the cost.

How easy is it for non-ML tech folks to create an ML deployment?
This is a very good question to ask and think about, as the answer brings out the ease of use of the ML tools and the infrastructure in place. This could also be used as a test that you conduct as an experiment every 6 months, where you request a tech (but non-ML) person to deploy something for you. Ask them to use the documentation and all the support channels that are usually available to an ML practitioner. Try to time how much time it would take them to do it, and try to improve on this time as you move ahead in the maturity of the ML system.

Are your ML deployments an easy target for security attacks?
How many times have you carried out pen testing on your ML deployments in last 6 months. All the folks that are hyped up on the MLSecOps zone would be super interested in this one, although I'm not gonna dive deeper into this for now. Having a good collaboration between the MLOps and security people can bring out a very solid security perimeter around ML servers as they are susceptible to one of the most creative attacks.

How granular is your isolation for ML deployments?
This is closely related to the blast radius of the deployments, as infrastructure provisioning should be done in a way that there's clear isolation between multiple deployments, also the usage of the service has to be defined in a way that one service being down does not' affect the performance of the other deployments at any layer.

Can users directly interact with your ML deployments?
One of the most important aspects of an ML deployment is who the deployment is going to be used by. IMO an ML deployment should only be consumed by another backend service, and never a human user. The issue is that due to the non-deterministic nature of ML services, added by the loose control over ML dependencies in python, its never a good idea to directly expose your ML server to a live user.

You'd always want to have a fallback when the model prediction doesn't make sense for any reason, you might as well build this fallback in the backend service directly interacting with the users and let the model servers only take care of the predictions. This also is inline with the single responsibility principle.[21]

How many times do your prod deployments crash in a month?
This is another metric based question, that is very important to track and use as a measure of the stability of ML systems of your organization.

What is the time taken from development to production?
The number of days it takes for a team on an average to take a model to production from the time active development has started post EDA, is a good metric to track and improve on.

Maintenance

Do you have runbooks that clearly define the next steps for on-call folks?
This directly relates to the existence of runbooks for alerts and issues that can pop-up and surprise you in production. Is there a clear process on when to involve ML folks, and is the process well known for all the people involved in the on-call process.

Do you have clearly defined runbooks when your central infrastructure is down?
Having runbooks for ML deployments is one simple chore, but making sure that there are runbooks available in the organization in the scenario of the central infrastructure being down is one of the most important runbooks to have.

How do you validate your production model predictions?
Every model deployment should have a clear definition of how the model predictions can be validated. A simple example could be a range value defining that the model predictions cannot be outside of.

---

[21] [21].

Do you have disaster recovery plans for ML?
Having a disaster recovery plan specifically for ML in place, and communicating it across should be a highly prioritized and frequently exercised plan at the organization.

Do you have a feedback loop for automated training in place?
One of the common problems for a lot of ML teams can be that they only train once, and the culture of iterating over the same model again and again as the need to update the model is missing in the culture as well as the priority of the project plans as well. If your ML practitioners don't think that their job is done until the model is either archived or retired, then the team is on a good path.

What is the time taken to detect a potential model drift?
Model drifts can be difficult to detect, depending on the kind of the problem the model is solving. Having a clear infrastructural setup in place that can store and alert about deviations in predictions of models and potential model drifts is very important. The time taken for the alert to trigger from the moment the model drift started could also be used as a metric.

Do you have clearly defined contracts and SLOs between teams?
Are there clearly defined and tracked SLO's that teams always strive to chase? These could help in making the interactions of the team very smooth w.r.t communications and responsibilities.

How big is your blast radius for ML Infrastructure?
Another one of my favorite questions, in case an ML server goes down, what is the expected blast zone it would affect? How many other model servers would it lead to not working? A good way to minimize this would be to make sure there are deterministic failover systems in place may one of the model servers not be working as expected.

People

How well do your *PMs understand the ML realm?*
One of the closest people that ML teams work with might be the project/product manager. A lot of times, the PMs can have unrealistic expectations about what is possible, how easy it is to make it possible, how long it would take to deliver an MVP, and most importantly, how important it is to work on the tech debt. Having a PM who has a well-versed understanding of machine learning can really make the development and maintenance an enjoyable experience for all the people involved. If you have a PM who used to be an ML practitioner, you can do wonders.

How mature is your infrastructure team in handling unexpected ML workloads?
This is a good way to understand how the ML culture is flowing through your organization. This would mean that incase there is no ML/MLOps person around, is the basic understanding of your core infrastructure team enough to be able to handle an issue with some ML infrastructure.

Do people in your organization understand what it takes w.r.t data, time, money and iterations to create a meaningful generative AI product feature?
I've been asked questions like *We gave you 20 examples this morning. Why don't we have a model yet that can classify stuff?* This question was asked to me in the afternoon of the same day. I hope no one ever has to go through such a situation, but a good measure to understand ML maturity is to check how well the non-ML folks in the organization understand Machine Learning.

Do people understand the difference between classical and ML software products?
There are a lot of differences between classical software products and ML, and the differences are much more than the non-deterministic nature of ML. Hoe many differences can a non-ML (but tech) person name if asked in your organization.

How well do collaborations between data and ML people happen?
The closeness of your data and ML team, can be a very big catalyst for getting cool and awesome ML products.

Do people in your organization think that ML is a magic stick?
If one can simply answer this question with a definite no, you are on a pretty solid grounds already.

Are people in the upper management familiar with the value and costs of creating and maintaining ML systems?
Most of the time, the people in the upper management of your organization are familiar with the value of ML, but when it comes to knowing the real cost in terms of efforts, time, uncertainty, and finances that go into developing a kick-ass ML product that delivers value to the users.

Metrics

These are some direct metrics that can be measured mathematically and can act as a very easy measure of how to understand the current state of MLOps and measure the changes that take place over a scale.

Average time *to production*

This is the average time taken by a team to get the first prediction from a modal from the time the project started. This measure could be a bit false in indicating the MLOps system, as sometimes the ML problem is complex enough that it would take a major chunk of the time taken, but the average should even it out a bit. This is still worth measuring and keeping in check.

Workflow failures in a month
This is the total number of workflows that fail in a month, whether it is due to infrastructural errors or the issues with the code being executed in the workflow component itself.

Time to try new things
This is the time taken to production for a project that is new to the organization and people. If there's a project that is different from all the other projects that the organization is used to or has carried out in the past, this metric is the time taken to take such a project to production.

Time taken to start EDA with new data
This is the time taken for a person to start working on EDA from the time they decided to use a specific data source.

Time to migrate to a new tool
Time was taken to completely migrate to a new tool X that aids the ML lifecycle by replacing an existing tool in use Y, from which it was decided that the Y needs to be and should be replaced by X.

Time taken to recover from disasters
Average time taken since the time an alert was raised to the time the ML system started performing normally with 100% of its capacity.

Time to react to incidents
The on-call team takes the time to start debugging the issue from the time the alert was raised.

Time taken to update a model in production
Time taken to update an already deployed model by a newer version from the time a person started working on creating a new model.

Incidents per month
Number of incidents that happen with ML services or tools.

Time taken per experiment iteration
Time taken for an ML practitioner to carry out an ML experiment at EDA or development time.

Ability to reuse resources
This could be measured as the moving average of reused components of a project by the total number of components of the project.

References

1. https://en.wikipedia.org/wiki/Test-driven_development
2. https://en.wikipedia.org/wiki/Directed_acyclic_graph
3. https://github.com/argoproj/argo-cd/tree/092bb7328cc7aeeb75460fc25fafa575418cacd3/docs/operator-manual/applicationset
4. https://en.wikipedia.org/wiki/Principle_of_least_privilege
5. https://github.com/argoproj/argo-cd
6. https://github.com/circleci/circleci-docs
7. https://github.com/fluxcd/flux2
8. https://github.com/pdm-project/pdm
9. https://github.com/python-poetry/poetry/
10. https://github.com/pypa/pipenv
11. https://github.com/pyenv/pyenv
12. https://github.com/ipython/ipykernel
13. https://github.com/mwouts/jupytext
14. https://github.com/jupyter-book/jupyter-book
15. https://github.com/wandb/wandb
16. https://github.com/aimhubio/aim
17. https://en.wikipedia.org/wiki/Wiki
18. https://en.wikipedia.org/wiki/Garbage_in,_garbage_out
19. https://en.wikipedia.org/wiki/Red_team
20. https://github.com/dependabot
21. https://en.wikipedia.org/wiki/Single-responsibility_principle

Conclusion

4

Emerging Trends

MLOps is a forever-evolving field, and new developments are happening every day. For example using Kubernetes for model deployment, the emergence of MLOps platforms, and the use of AI to automate parts of the MLOps process. As we move towards more and cheaper usage of Generative AI in our daily lives, the MLOps landscape might look completely different than what it is today. But let's discuss what are some notable moments and trends that have driven the field uptill now.

The use of kubernetes as the default choice of infrastructure, has been one of the most foundational changes that has led to most of the tools being written in cloud native choice of supported compute. Most of the well known MLOps tools that have been discussed in this book are cloud native and use kubernetes as their choice of infrastructure.

It's not just the mlops pipeline tools, but most of the tools that provide serving layers or ML model serving frameworks are also usually very tightly coupled with the Kubernetes cluster, the service, and network management of the cluster.

Along with the heavy use of Kubernetes, the change of DevOps roles to get used to ML workloads is another new trend that can be seen commonly. A lot of infrastructure folks are coming closer to ML by hopping on to the MLOps responsibilities. Some ML engineers are moving towards MLOps work by getting used to infrastructure and understanding the complexities of scale. The SRE handbook[1] can act as a really good resource for people trying to come closer to infrastructure.

It's also very useful to automate parts of you ML project using generative AI. Processes like pre-processing or feature extraction have known to be automated by sending a well designed system and user prompt to an large language model. There are more

---

[1] [1].

such stages like writing infrastructure of kubernetes manifests, that can be automated completely using GenAI.

Apart from the core MLOps work that has effect from ML developments, the number of MLOps platforms that have flooded the market is huge. There are a lot of startups offering whether a complete, one platform solves all solutions or solving a concrete niche problem with a good solution. One can often find oneself in a conundrum when trying to understand which tool would work for them, but make sure you stick to the definition of your needs and evaluate tools until you are really confident that the tool solves your needs.

Another interesting emerging trend is related to the issues that Python brings to the table. As we move towards more and more complex software engineering needed for ML models, everyone is realising that Python is not the best language when it comes to handling dependencies. This has led to the recent rise of people switching to rust[2] or other similar options that they can explore. This moment is still not hyped a lot, but there has been a drive to move towards either better dependency management (check out nix[3]) or changing the default language that ML is written in.

While talking about emerging trends, we cannot leave RAGs and langchain, RAGs, or retrieval augmented generated (as the name suggests) allows you to send in an additional context that is used to generate answers. In this method, your model generates content by retrieving it from the known set of contexts provided to it. This really helps narrow down the scope of possible answers to specific contexts, inherently increasing the precision and accuracy of the content being generated as the answer.

LangChain, on similar notes, has been really helpful in letting people build applications on top of existing language models by allowing people to use multiple steps along the line. Databases are used to support memory and context retention across a large timeline and scale. It simplifies the integration of these language models with various data sources and offers a lot of other tools that help build useful functionalities on top of the existing models available. The focus is to build chains of tasks that can be run in tasks that can provide a modular interface for people to build on easily.

With these kinds of developments in the ML space, it would be really interesting to see how the field matures and how it affects other software related fields that exist today. I might as well go ahead and say that I see a future where ML is not going to be something special. We are gonna use ML so much so in the development of any software product or service that ML will be a default prerequisite for any software development as a standard set of skills that everyone should possess.

---

[2] [2].

[3] [3].

Collaboration and Communication

In the world of ML and data science, collaboration isn't just nice to have. It's actually a required fuel to keep everything moving forward. We might have top-notch data scientists and engineers, all experts in their own areas, but in MLOps, it's the combined effort of the whole team that really makes a difference. And it's not just about making accurate models. It's about building them to be scalable, efficient, and maintainable. So, you have to have strong communication and collaboration to make sure everyone's working toward the same goal, and there is no room for confusion. This will enable the team members to share insights freely, make people comfortable with bringing up potential issues, and keep everyone aligned throughout the development and deployment phases.

When it comes to a smooth MLOps process, having clear communication is almost half of the battle. It's important to set up clear roles and responsibilities for everyone (data scientists, software engineers, DevOps) involved in getting models into production. The contracts will help prevent any misunderstandings and confusion that can come up between teams and people. For example, data scientists need to specify what data they need and how it should be preprocessed, while developers have to explain how they'll integrate the model into production. DevOps folks are in charge of building the deployment pipeline, and the operations team needs to make sure everything's running smoothly once deployed.

An MLOps engineer kinda sits at the intersection of data, ML, and site reliability engineering, and because of that, communication is pretty much one of the biggest parts of the role. When everyone's clear on their roles, things just go smoother. Data engineers can focus on scaling data pipelines, data scientists can work on refining the models, and DevOps engineers can focus on setting up deployment infrastructure—all of this together makes the delivery and maintenance of ML models that much easier.

Setting up roles is one thing, but it's just as important to keep communication channels open. Regular meeting and collaborations make it super easy to track progress and bring up any blockers. When people have clear expectations, you can understand the journey, estimate timelines, track progress, spot issues early, and make the necessary tweaks wherever needed. It prevents delays and keep the deployment process as smooth as possible.

Conclusion

As I wrap up this book, I hope it took me through a journey of describing various aspects of implementing machine learning operations in the real world. I hope the third chapter keeps giving you challenges to chase and acts as the north star for you to chase. If I have to summarise a few key takeaways, I would say:

- Listen to the pains of the Machine learning engineers. Developer experience matters the most.
- Make sure you never compromise on the quality of the models that are shipped to production, even if this comes at the cost of developer experience for some time.
- Production environments are sacred and should be treated so.
- Embrace a collaborative mindset, and be ready to solve hard problems.
- No solution is perfect; be ready to iterate over and over again.

As this filed is ever evolving, and the content of this book is just the beginning. As new challenges emerge, one should stay curious and acceptable to change, while having the user (ML practitioner) in mind. I hope this book serves as the foundational launch bed into this field.

MLOps isn't just about tools; it's about processes, mindset, and building systems that scale.

References

1. https://sre.google/sre-book/table-of-contents/
2. https://github.com/rust-lang/rust
3. https://github.com/NixOS/nix